Management Principles and Practice

A Cybernetic Approach

Studies in Cybernetics

Edited by F.H. GEORGE, Brunel University

A series of books and monographs covering all aspects of cybernetics.

Volume 1
INTRODUCTORY READINGS IN EXPERT SYSTEMS
Edited by D. Michie
University of Edinburgh

Volume 2
A CYBERNETIC APPROACH TO COLOUR PERCEPTION
N.C. Paritsis and D.J. Stewart
Brunel University

Volume 3
MANAGEMENT PRINCIPLES AND PRACTICE
A Cybernetic Approach
R.H.D. Strank
Brunel University

Other volumes in preparation

ISSN: 0275-5807

Management Principles and Practice

A Cybernetic Approach

R.H.D. STRANK
Brunel University

GORDON AND BREACH SCIENCE PUBLISHERS
New York   London   Paris

Gordon and Breach, Science Publishers, Inc.
One Park Avenue
New York, NY 10016

Gordon and Breach Science Publishers Ltd.
42 William IV Street
London, WC2N 4DE

Gordon & Breach
58, rue Lhomond
75005 Paris

Library of Congress Cataloging in Publication Data
Strank, D., 1940-
    Management Principles and Practice: A Cybernetic Approach
    (Studies In Cybernetics: v.3)
    Bibliography: P.
    Includes Index.
    1. Organization. 2. Management. 3. Cybernetics.
I. Title. II. Series.
HD31. S696388   658   81-13263
ISBN 0-677-05850-0   AACR2

For Margaret

Contents

Introduction to the Series   ix
Preface   xi
Acknowledgements   xiii

1. The Importance of Organisations   1

2. The Nature of Organisations   4
2.1 Introduction   4
2.2 The Structure of Organisations   7
2.3 The Dynamics of Organisations   21
2.4 The Control of Organisations   27
2.5 People in Organisations   31
2.6 The Environment of the Organisation   34
2.7 Summary   36

3. The Cybernetic View of Organisations   38
3.1 A General Overview   38
3.2 Feedback and Anticipatory Control   41
3.3 Neural Nets and Entropy   45

4. The Cybernetics of Management   52
4.1 A Model of Organisation and Management   52
4.2 Internal Regulation   54
4.3 External Regulation   83
4.4 Summary   85

5. Some Consequences of the Model   88
5.1 Introduction   88
5.2 A Practical Study   88
5.3 Some Theoretical Aspects   99
5.4 In Conclusion   138

References   139
Index   143

Introduction to the Series

Cybernetics, the science of systems of control and communications, is a rapidly growing subject and there now exists a vast amount of information on all aspects of this broad-based discipline. To call cybernetics 'broad-based' is to imply that its viewpoint is nearly identical with all the approaches taken to artificial intelligence. Furthermore, systems analysis, systems theory and operational research often have a great deal in common with cybernetics – and, in fact, are not always discernibly different from it, so far as this series is concerned. Computer science, too, is usually closely linked.

The fields of application are virtually unlimited and applications can occur in investigating or modelling any complex system. The most obvious applications have been to construct artificially intelligent systems to simulate the brain and nervous system, and social and economic systems.

The range of applications today has gone so far from its starting point that it now includes such subjects as aesthetics, history and architecture. The immediate modelling can be carried out by computer programs, special purpose models (analog, mathematical, statistical, etc.), and automata of various kinds, especially neural nets. All that is required of the system to be studied is that it is complex, dynamic, capable of 'learning' and has feedback or feedforward or both.

This is an international series. It includes translations in English from originals in other languages.

FRANK GEORGE

Preface

The origins of this work go back several years to when I first became interested in management as a topic for study in its own right. One of the first things I did was to look for a basic theoretical framework which would explain what management was for, and the various ways in which managers functioned.

Rather to my surprise, I could not find it. There was, to be sure, a very great deal written, but it was basically descriptive, giving an account of what went on. There was little that sought to provide a generalised, scientifically-based, theory of how and why organisations function. Many writers had explored particular facets of the subject, but from different viewpoints, and there was nothing that put these different aspects into relation with one another.

This book sets out to show that cybernetics can supply this basic theoretical framework, which allows management to be studied as a phenomenon in its own right, and to provide greater insight and understanding of its processes and functions. It follows that it should be of interest not only to cyberneticians but also to those concerned with understanding organisations from a practical viewpoint.

Different parts of it will probably be of interest to these different groups. The first part, which deals with the nature of organisations, presents little that will be new to the serious student of management, but may well be of interest to the cybernetician who is new to this field. Conversely, the second part will perhaps be of more interest to students of management than to cyberneticians; it takes some very basic cybernetic concepts and shows how they can be used to build a detailed and comprehensive model of management and organisation. The third part, which develops some consequences of this model, should be of interest to both, though for different reasons. From the practical viewpoint, this part shows how cybernetic ideas can lead to better understanding of management, and to improved operational effectiveness. From a theoretical viewpoint, some of the concepts that emerge have wide and important applicability.

Overall, it is hoped that this book shows two things. Firstly, that cybernetics is a wider discipline than it sometimes may appear, especially to an outsider, that it is not necessarily limited to topics such as Artificial Intelligence or Computer Science. Secondly, that management can benefit from applying comparatively simple analyses to actual situations. Only the outermost fringe of this latter aspect has been touched on; there remains an enormous amount to be done.

R.H.D. STRANK

Acknowledgements

If I were to try to make a tally of all the people who have helped, supported and encouraged me in their various ways, the resulting list would resemble nothing so much as a telephone directory. It is in the nature of things, however, that certain entries stand out from the rest. Their names will have to stand on behalf of many.

First my thanks must go to Gordon Cutcliffe. His interest and encouragement during the early stages of this work (now many years ago), and especially his development of it, constitute a debt I am only too pleased to acknowledge.

Special mention must also be made of Professor Frank George and Dr David J. Stewart of Brunel University, whose efforts to correct my errors have, I hope, not been entirely unsuccessful. Nor would I wish to overlook the patience of the ever-helpful Mrs Kilbride.

But my final word of thanks must go to my wife and family, without whose tolerance and understanding during long hours of writing, this book would have been impossible.

1

The Importance of Organisations

Organisations of one form or another have been a feature of society for a long time. Power and authority have been exercised through formal institutions since the beginnings of recorded history. What is more, organisation seems to be a characteristic trait of human society, and fundamental to it.

Certainly, organisations are a striking feature of present-day life, to the point where the overwhelming majority of human effort is channelled by organisations of one kind or another, whether these be business firms, government bodies, charities, clubs, or others. Furthermore, they are tending to become larger, both in terms of their scale of operation and the numbers of people involved, and thus are becoming more complex entities. Examples of this can be seen in the growing band of international business enterprises, and of supranational government bodies such as the UNO, OPEC, the EEC and many others.

The proper functioning of organisations is thus a major interest of civilised society. It is readily apparent that all is not well in the institutions we have at present. There are many signs of dissatisfaction with, and alienation from, present organisations (not that this is a particularly new phenomenon). In the political sphere, revolutions, coups, and terrorism are reported on an almost daily basis. In the business sphere, strikes, disputes, and poor productivity are accepted facts of industrial life in many economies. All in all, there is considerable room for improvement.

The aim of this book is to make a contribution to the understanding of organisations. Since many other people have written in this field, it would be as well to start by trying to explain what it is that is being attempted, and how it differs from much other work.

This book seeks to set out a theoretical basis for the scientific study of organisations, and in particular for the scientific study of the structuring and functioning of management activities. It does not attempt to provide instant answers to the problems that beset managers from day to day. The aim is much more to supply a broad analytical framework within which management activity can be assessed, quantified, and studied. This is done in the belief that major improvements in our institutions will only arise from a much better understanding of how they actually work, and the best way to achieve this is through a process of scientific study.

It follows naturally that this book is not normative — it does not try to prescribe what should be done, how organisations should be structured, how management should function. The purpose is to develop a model of organisational functioning that is sufficiently general to encompass a wide variety of styles and methods, yet sufficiently precise to enable it to be used for the study and assessment of managerial work. How far it is successful, the reader himself may judge.

This book falls into three main parts. The first describes the nature of organisations, what they are for, what they do, the way they behave, and the problems they encounter. This is done through a survey of the literature, which is necessarily representative rather than comprehensive, and seeks to identify the dominant themes in management thought and to offer some initial evaluation of these ideas. The chief purpose of this section, however, is to present what may be thought of as the basic facts which any theory of organisation and management must be able to explain.

The second part surveys the contribution of formal scientific study to the field of management. In doing this, it is important to draw a distinction between the application of scientific method to the solution of operational problems faced by managers, as distinct from the study of the processes by which organisations are managed. The former is the province of what may generically be termed the 'management sciences' — such as OR, work study, quality control and the like — and is not the concern here. It is the latter aspect — the issues of structure and function — which is intended to be examined, and which has been the subject of comparatively little attention. Such work as has been done has been carried out in the field of cybernetics or closely allied disciplines, and it is this branch of science that has been drawn on to supply the basic concepts used in the third part of the book.

This third part develops a model of organisation and management. In doing so it draws on previous work, but extends this to a greater level of detail and also to cover a more comprehensive range of organisational activity than other studies. It also discusses how the model is able to account for the reported features of organisational behaviour, at least in a broad level of detail. Additionally, although the treatment is basically non-mathematical, some indications are given of the type of mathematics that appear to be appropriate for the further development of the approach.

The fourth and final part looks at some of the implications of the model. Firstly, to show how it can be applied to practical problems, a study of an actual managerial situation is reported. Following on this, some of the theoretical consequences of the model are examined to try to gain a deeper understanding of the role of the manager.

This need for greater understanding is, in fact, the theme and mainspring of the book as a whole. To reiterate a statement made earlier, all is not well with our organisations and there is a need to improve them. It would be a great mistake, of course, to think that science by itself holds the answer; there are other issues involved. Nevertheless, it has its proper place and its own contribution to make.

In this context, it is not crucially important whether the scheme put forward here is precisely accurate in all details. What is important is that it offers an account of organisation that can be tested by experiment and, thus modified, improved or discarded as found necessary in the light of experience. In this process, much will be learnt.

2

The Nature of Organisations

2.1 Introduction

Organisations are so familiar in everyday life that their nature and behaviour may well be taken for granted, with little or no thought as to their characteristics. It is perhaps as well therefore to start with an examination of what organisations are.

A formal definition sufficient for our purposes has been given by Barnard (1948) as:

A system of consciously co-ordinated activities or forces of two or more persons.

This is an extremely wide-ranging definition, sufficient to encompass the State, Trade Unions, religious bodies, industrial companies, and charities.

Other workers have provided essentially similar views. For example, Bakke (1959) sees an organisation as "... a continuing system of differentiated and co-ordinated human activities which welds together resources into a whole that has a character all of its own". The concept can be traced back to Aristotle who wrote, "Men journey together with a view to general advantage, and by way of providing some particular thing needed for the purpose of life . . . ."

Some aspects of Barnard's definition deserve comment. Firstly, it implies that the essential component of all organisations is a group of people and therefore organisation is an essentially human activity; it is worth quoting a further remark of Bernard (*ibid*) on this topic namely that "... a co-operative system is incessantly dynamic, a process of continued readjustment to physical, biological and social environments as a whole". Not only does this encapsulate the essential nature of organised activity, it suggests powerfully that the disciplines of cybernetics, which are accustomed to treating complex dynamic systems, are appropriate tools to investigate the problems of organisation.

A second point about Barnard's definition is that organisations are characterised by shared tasks and hence, since tasks can always be construed

4

as having a purpose, by purposes held in common by the group. A pedantic point here is that it is not necessary that the task (or tasks) are beyond the capacity of a single individual to accomplish. Thus although it is quite possible for a man to build his house through his own efforts, it is more common to find that houses are built by an organisation. A further observation, which Barnard himself makes, is that there is a sense in which the tasks of an organisation are quite specific (e.g. "build *this* house", "cash *this* cheque") and its purpose is accomplished when the task is completed. Thus, in carrying out its tasks, an organisation accomplishes its purpose and, logically, should disband itself. For an organisation to continue, it therefore needs to adopt new objectives continually.

Occasionally, this process can be observed in action. A recent example has been the activities of CAMRA (the Campaign for Real Ale). Formed originally to promote the availability of particular types of beer in public houses, it was largely succesful in this aim. It then moved on to other (though related) activities, actually running public houses, and also became more involved in political issues such as trading monopolies. More recently, it has started to consider brewing its own beer.

It is more general, however, for organisations to overcome this dilemma by adopting a statement of purpose at a more generalised, abstract level such as "to make motor cars" or "to provide a banking service", which allows fresh tasks (and hence purposes) to be generated on the completion of a given task. This is an important point from a philosophic point of view, particularly when discussing the objectives of an organisation. It implies that organisational objectives are not fixed for all time, but are themselves evolving as part of the ". . . process of continued readjustment" referred to above. It also offers an explanation of why many workers in the field of organisation, particularly of business management, find that definition of objectives is a recurring theme. Grainger (1964) goes so far as saying that objectives ". . . should be period-ically reconsidered and redefined, not only to take account of changing conditions, but for the salutary effect of rethinking the aims of organisation activities". In similar vein, Humble (1968) has written, "It is always stimulating and constructive to look afresh and critically at the company's forward plans, particularly as the range of objectives is often found to be dangerously restricted."

Yet another aspect of the topic of organisational objectives, which again is recognised by Barnard, is that not all individuals within an organisation will be fully committed to them — in fact some may be opposed to them. Further-more, this degree of committment may be expected to vary through time.

A quite separate aspect of the definition of an organisation is the emphasis that it places on "consciously co-ordinated activities". This is a key feature, one that differentiates an organisation from a mob or a haphazard, accidental collection of individuals. Additionally, it brings into focus the clear need for a means of dividing work between individuals, and a mechanism of communication and control to achieve this co-ordination. This aspect is of such primary importance to the success or otherwise of an organisation that the term "organisation" itself is frequently used to denote exactly this, i.e. methods by which work can be divided up and subsequently controlled. The word is then used as an abstract, rather than a concrete noun.

Most writers in this field tend to use the term "organisation" in this more abstract sense, and comment on the nature of "organisation" from a variety of points of view. There is such a wide range of material published under this general heading that it is not practical here to review it all in depth. However, it is possible to pick out some of the main strands of thought and progress, each associated with a particular group or school of individuals.

As with most topics, it is possible to trace discussions of organisation back to very early times. For example, Plato makes reference to the organisation of the State in *The Republic*, particularly in Books II and VIII and says much which is still of relevance today. However, modern approaches to organisation can be seen to start to emerge at about the beginning of this century, and it is convenient to review it under five main thematic headings. Before doing so, it should be pointed out that most of the work referred to deals more or less explicit with industrial and or business activities. Whether these are an appropriate model for other types of organisation is a question which is touched on below.

The five main themes which can be discerned are:

1) The structure of organisations.

2) The dynamics of organisations.

3) The control of organisations.

4) People in organisations.

5) The environment of the organisation.

It will be appreciated that this set of headings is to some extent arbitrary. Many others could be devised. Nor is it an entirely satisfactory set, for there is much overlap between the different areas and one blends into another with no very clear dividing line. Their chief merit is that of convenience, in being able to bring together a number of different contributions. They are not independent factors in organisation but rather different aspects of the whole which need to be considered together but may be conveniently discussed separately.

2.2 The Structure of Organisations

It is everyday experience that organisations have a hierarchy, the boss at the top, the workers at the bottom, and a number of levels in between.

It is not meant to suggest that this is the only possible form that organisations could take, but merely to point out that it is a very common, if not ubiquitous, feature of bodies that actually exist. Along with this hierarchy go the concepts of specialisation, task allocation, supervision, communication, co-ordination and the like. Thus the hierarchy has a structure within it which defines the organisation. As Cameron (1948) puts it, this structure is "The framework of duties and responsibilities through which an undertaking works."

There is, of course, no one unique way to divide tasks. Equally, there are few enterprises that carry out *exactly* the same total function as another, even though they may do many similar things.Thus it can be expected that each organisation will have a hierarchy that is, in some respects at least, unique to itself. Indeed, it is quite possible that two companies carrying out broadly similar activities may have completely different structures. Nevertheless, several writers have investigated organisational hierarchies to see if there are any general forms or features that can be extracted.

One of the early workers in this field was Weber (1930, 1947) whose prime concern was to postulate classifications of organisation structure, particularly in relation to the authority structures within them. Perhaps his principal contribution was his analysis of the basis on which one person exercised authority over another. He distinguished three main principles which he labelled "charismatic", "traditional" and "rational-legal" — which last has subsequently been relabelled as "bureaucracy".

"Charismatic" is a term which can be translated as "leadership", the quality or qualities that enable one man to inspire others to do as he wishes. Organisations based on this type of authority do exist, but, as Weber points out, they tend to be unstable. Once the charismatic figure passes on, or loses his charisma, the basis of authority has gone. (Religious organisations are an interesting apparent exception to this rule.) The organisation then needs to substitute some other form of authority, or it falls apart.

Weber's "traditional" organisation overcomes this problem by granting authority on the basis of precedent and usage. Weber drew upon mainly historical illustrations for this type of organisation, particularly feudal systems, but examples can still be found in modern society — it is not unknown for instance for promotion to senior executive positions in business to be the result of being related to the Chairman of the Board. Equally, membership of committees can be granted on the basis of holding a certain position — membership "ex-officio".

Weber's third type, which he termed "rational-legal" comes closest to current concepts of organisation. Authority within bureaucracy (as this category has been renamed) is exercised through an accepted system of rules and procedures, and individual authority derives from the role or office which a person holds. In current usage, the term bureaucracy has become synonymous with hidebound, over-formalised, inefficiency but this is not how Weber originally conceived it. In his view, "The decisive reason for the advance of bureaucratic organisation has always been its purely technical superiority over any other form of organisation", because it is devised specifically for the purpose for which it is intended.

It is difficult to quarrel with this conclusion in the form in which it is stated, simply because his definition of bureaucracy is sufficient to include any form of structured, task-oriented, behaviour — i.e. any form of organisation. It is also fairly clear that Weber's three types are not mutually exclusive categories, and all three may co-exist in any given organisation at a given time.

Nevertheless, Weber made an important contribution, in that his was the first attempt to produce any organisational categories at all. Furthermore his categorisation gives some insight into an important aspect of organisational behaviour, the use of power and authority. Other work has followed on from his lead, such as that of Gouldner (1955), who expands on Weber's original single concept of bureaucracy and identifies three sub-classes, "mock", "representative" and "punishment-centred".

In a "mock" bureaucracy the rules are imposed by some outside source, rather than derived from the nature of the task and the authority-structure within the group; for example regulations imposed by public authorities such as the Factory Inspectorate. "Representative Bureaucracy" is much closer to Weber's original concept; rules are promulgated by "experts", whose authority is acceptable to all the members of the organisation. The rules are accepted by both superiors and subordinates, because they derive from values held in common. "Punishment-centred" bureaucracy, arises when values are not held in common, and rules derive from the efforts of pressure groups (which may be management or workers) to enforce their will on other groups. Deviations from the rules are punished by the pressure group concerned.

As analytic tools these categories also suffer from the fact that they are not mutually exclusive, and can co-exist in one group. Indeed, Gouldner's prime use of them was to study a situation where the organisation changed from one pattern to another, and to explain the tensions and disruptions that occurred within this framework of categories.

A quite different approach to organisational structure is presented in the work of Woodward (1958). This is an empirical study of organisation structures found in practice, covering 100 firms of medium-large size in south-east Essex. The variables in the study included the number of levels of authority, the span of control (i.e. the number of direct subordinates reporting to a superior), the degree to which duties were defined, amounts of written communication, and the use of specialisation. She attempted to relate these variables to the types of technology and production system used. Many relationships emerged, among the more significant being that the number of levels of authority increased with the technical complexity of the process. She also observed that difficulties were generated when (due to a takeover) it was attempted to replace an organisation suited to one scale of production with one applicable to a larger scale.

However, the main conclusion that Woodward drew from the many relationships she examined was there is no one best form of organisation. Organisation, she says, should be adapted to the demands imposed by the objectives and technology of the individual firm. Whilst this is not a conclusion to be contested at this point, it is difficult to see upon exactly what grounds Woodward bases it; she included no criteria of organisational effectiveness in her study. As far as can be gathered, the conclusion depends on the assumption that the firms in the survey had adopted the best form of organisation for their needs.

A further difficulty in interpreting Woodward's work is that it is comparatively narrowly based, in the sense that it was concerned only with manufacturing organisations. Commercial, or marketing, aspects were not included let alone non-business organisations.

Support for Woodward's main thesis can be found, amongst other places, in the work of Burns and Stalker (1961). They came to essentially the same conclusion through starting from a different, basically sociological, viewpoint. Their studies were again concerned with manufacturing industry, and particularly with the problems associated with major technological innovation. They came to the view that organisations can be categorised along a continuum the end points of which they called "organic" (or "organismic") and "mechanistic". The "mechanistic" type, which in many ways corresponds with Weber's bureaucracy, is characterised by clearly defined vertical hierarchies of command, with the overall task divided into specialisms. Tasks for individuals are carefully set out in detail, and great emphasis is laid on adherence to rules and procedures. The "organic" type is characterised by a much more flexible, informal, system where individuals' tasks are apt to be continually changing, dependant upon the nature of the problem of the moment. There is much greater emphasis on horizontal communication and interaction and correspondingly less on formal channels and formal authority.

Burns and Stalker relate these types of organisation to the stability of the conditions in which the organisation is working. "Mechanistic" organisations, they argue, are adapted to relatively stable conditions, whereas the "organic" type is adapted to unstable situations where new problems arise frequently, problems which cannot be slotted into an existing specialist role for a solution.

Again, this study can be criticised on the grounds that the criteria for an effective organisation are ill-defined. There is little attempt to assess the quality of management that was operating within the various structures described, and there is no attempt to disentangle the effects of this variable. However, the study is valuable in that it demonstrates something of the wide variety of structures found in practice, and provides a further dimension for the analysis and understanding of organisation. Furthermore, it is interesting to compare this work with that of Emery and Trist (1960). They report the results obtained with two different types of organisations working the same technological process. The types of organisation they classify as "conventional" and "composite" which appear to be similar in all essentials to the "mechanistic" and "organic" categories respectively of Burns and Stalker. They found that efficiency, in terms of variables such as output, hours worked, breakdowns, was significantly influenced by the type of organisation structure. Two cases were reported, one of coal-mining, one of weaving. The coal-mining study showed that the "composite" system was superior, and the authors comment that the task was complex due to the constantly changing underground conditions. In contrast (and Emery and Trist do not appear to have realised this) the weaving study showed superior performance with a much more "conventional" structure. This may have been related either to the more predictable nature of a weaving task, or to the level of technical skill and comprehension amongst the operatives — the weaving study was carried out in India.

Studies such as these show something of the complexity of the structure of organisations. However, the predominant strand of managerial thinking on the structure of organisations has its origins in the work of Fayol (1908). He wrote from theoretical interest or experimental observation, but nevertheless his work has gained a wide and enduring reputation. He enunciated 14 "principles of management", several of which are concerned with organisational structure. Those most relevant for the immediate purpose here are as follows (using Fayol's original numbering):

1) *Division of Work*. This, of course, is the basis of all organisational activity, although Fayol does not specifically say so. He sees the point of division of work is to increase efficiency ". . . to produce more and better work with the same effort", and he sees it as a principle applicable to work of all kinds, not simply manufacturing. Interestingly enough, he seems to have been aware that specialisation of work can be carried to excess. He says, ". . . yet division of work has its limits which experience and a sense of proportion

teach us may not be exceeded."

2) *Authority and Responsibility.* Fayol distinguishes two types of authority, one derived from personal qualities, one derived from official position. (He makes no mention of Weber's third source of authority, the "traditional".) He sees an important aspect of a good manager as the fusion of these two types in one individual. Equally, he is insistent that authority and responsibility are co-extensive.

3) *Discipline.* Fayol distinguishes this quite clearly from authority. He defines it as follows: "Discipline is in essence obedience, application, energy, behaviour and outward marks of respect observed in accordance with the standing agreements between a firm and its employees . . . ." It is clear that he views discipline as operating within a set of (more or less) formally defined rules and procedures, and that discipline applies as much to managers as subordinates. Discipline should be exerted on an agreed basis, fair to all parties, and includes the use of sanctions where it is breached.

4) *Unity of Command.* This is perhaps the most fundamental of Fayol's principles of organisation structure. From it flows naturally the whole concept of the hierarchy of command and the typical pyramid structure of management. In simple terms, "unity of command" can be expressed as "one man, one boss", which is an exact paraphrase of Fayol's words, "For any action whatsoever, an employee should receive orders from one superior only." He also says, "This rule seems fundamental to me and so I have given it the rank of principle." It is clear that he recognised that his principle was not universally observed, and he illustrates some of the situations that arise when it is not.   –

5) *Unity of Direction.* This is an extension of the "unity of command". It is defined as " . . . one head and one plan for a group of activities having the same objective". Unfortunately, Fayol does not make clear how it is to be established which activities have a common objective; his statement can be interpreted in at least two senses, one product-oriented (i.e. to produce and sell a given article or service) one process-oriented (i.e. to produce a range of articles or services). This is a theme in organisation structure which has received much discussion, and is still not resolved. Indeed, it seems that the question may never be answered, but resolved through progress to new types of organisation structure (see, for example, Newman (1973)).

8) *Centralisation.* This is still very much a problem in current organisational design, and it is of interest that Fayol identified it so long ago. He defines it as follows, "Everything which goes to increase the importance of a subordinate's role is decentralisation, everything which goes to decrease it is centralisation." He also comments that the issue of centralisation or decentralisation is one of degree, not of principle. He interprets it in terms of

the length of the "Scalar chain" (see below), and as being dependant upon the abilities and disposition of the managers involved.

9) *Scalar Chain*. This is the line of formal authority from the lowest operative to the highest authority, and is essentially an interpretation of the principles of unity of command and unity of direction into their hierarchical consequences. Fayol uses this to discuss the need that can arise to short-circuit the normal channels of communication. He apparently feels that communication within organisations should be basically "vertical" and that "horizontal" communication should be resorted to only in emergency.

Of Fayol's 14 principles, the foregoing are those most directly concerned with the structure of organisations. The balance are concerned more with the functioning of organisations, though the distinction is not always easy to draw. They are worth quoting because they form the foundation of a great deal of subsequent work. Furthermore, little of fundamental importance has been added to Fayol's principles, although they have been refined, reshaped, and reworded. This is not to say that there is a general concensus that Fayol's conclusions were correct, but rather that he identified with clarity the major issues to be resolved in structuring an organisation. The debate on their correct solution still continues.

Contemporary with Fayol was Taylor, (1903), who founded the Scientific Management movement. However, he contributed little to the theory of the structure of organisation; many of the principles generally accredited to him were in fact originated by Fayol. Taylor's chief contribution in this area (which is overshadowed by his contributions in other areas) was his concept of "functional management", particularly the "functional foreman". Under this scheme, every worker had several foremen in charge of him, each responsible for a specific aspect of performance, such as discipline, speed, and quality. Although this concept did not enjoy a long application in practice, it did serve to introduce the notion of "functionalism" into the analysis of organisations, where it has remained.

Several writers have taken up the themes initiated by Fayol and Taylor, among them Sheldon (1924), Lee (1925), Robinson (1925), Mooney and Riley (1931). Their views were synthesised in the work of Urwick, who has written widely on the subject of organisation and management. His views developed over the years, and perhaps the definitive statement of them can be found in his *Notes on the Theory of Organisation,* published in 1952. In this, he identifies eight principles of organisation, as follows:

1) *The Principle of the Objective*. Every organisation, and every part of the organisation, must be an expression of the purpose of the undertaking concerned or it is meaningless and therefore redundant.

2) *The Principle of Specialisation.* The activities of every member of an organised group should be confined, as far as possible, to the performance of a single function.

3) *The Principle of Co-ordination.* The purpose of organising, *per se,* as distinguished from the purpose of the undertaking, is to facilitate co-ordination, unity of effort.

4) *The Principle of Authority.* In every organised group the supreme authority must rest somewhere. There should be a clear line of authority from the supreme authority to every individual in the group.

5) *The Principle of Responsibility.* The responsibility of the superior for the acts of his subordinate is absolute.

6) *The Principle of Definition.* The content of each position, both the duties involved, the authority and responsibility contemplated, and the relationships with other positions, should be clearly defined in writing and published to all concerned.

7) *The Principle of Correspondence.* In every position the responsibility and the authority should correspond.

8) *The Span of Control.* No person should supervise more than five, or at the most, six, direct subordinates whose work interlocks.

9) *The Principle of Balance.* It is essential that the various units of an organisation should be kept in balance.

10) *The Principle of Continuity.* Reorganisation is a continuous process; in every undertaking specific provisions should be made for it.

The work of Urwick represents the conventional wisdom of managerial views on organisation structure. For that reason, these principles are worth some review.

The first point to be made is that they accept implicitly an authoritarian and hierarchical structure. The possibility of any other form of organisation is not even admitted, let alone discussed, and the line of descent from Weber's bureaucracy through Fayol and Taylor is clear. Rather than principles of organisation they are perhaps best viewed as a summary of the characteristics of one particular dominant form, essentially Weber's "rational-legal" system or Burn's "mechanistic" type.

From a more philosophic viewpoint, Urwick's principles are bedevilled by lack of definition of terms. Thus, The Principle of the Objective founders on the problem of defining an organisation's objective, as discussed above. This is particularly so when one attempts to discern an overall objective through a review of an organisation's activities, for then by definition, ". . . every part of the organisation must be an expression of the purpose of the undertaking. . ."

In a situation where objectives are bound to be underspecified, the use of Urwick's first principle as a tool of organisation design is extremely limited.

Similarly, the use of the second principle depends upon being able to specify exactly what activities constitute a function. Since a function is an abstract concept that can be built up to any desired level of generality, it is difficult to see how the work of an individual can fail to be "confined to the performance of a single function", given adequate ingenuity in finding the appropriate descriptive phrase. Thus, the usefulness of the second principle is open to doubt.

The third principle, that of "co-ordination" is perhaps unexceptional in itself as an expression of good intent. Again, however, it is of little practical use as a guide when actually designing an organisation.

The Principle of Authority is a statement about the nature of hierarchies, and does not greatly advance understanding of this subject. Furthermore, as stated, it does not give any lead as to where one might expect to find the ultimate authority nor whether it rests with one individual or a group. Equally, it does not explicitly acknowledge Fayol's Unity of Command, though one must assume this is through oversight rather than intent.

Urwick's fifth, sixth and seventh principles do not seem relevant to the topic of organisation structure. They are much more concerned with managerial practice within a structure.

The eighth principle, that of "Span of Control" is the one statement that is directly and practically useful in organisation design. It is interesting that Urwick does not dignify it with the title of "principle". Whether it is a reliable guide in practice is more doubtful, for it is built on a rather dubious base. Two sources can be traced for his statement, the first in the work of Lee (1925) as an empirical observation,

It seems from practical experience that in no case should a manager have more than five representatives of divisions in touch with him, whether these divisions are what one may call territorial, functional or technical.

The second source is in the work of Graicunas (1933). In essence, his conclusion was based on the number of permutations possible amongst a given number of people, and the rapid increase in this number of permutations as the size of group increases. He expressed this as the number of "relationships" within a group. (A summary of his results is given in Table 2.1 which is a greatly simplified version of Graicunas' original.)

He then invokes the psychological notion of "span of attention", without quantifying it, and states that his opinion is that 222 relationships (= 6 subordinates) is about the maximum that any individual should be expected to enter into. He also notes that the rise when a seventh person was introduced (to 490 relationships) was considerable.

TABLE 2.1

No. of subordinates	1	2	3	4	5	6	7	8	9	10	11	12
No. of relationships	1	6	18	44	100	222	490	1080	2376	5210	11374	24708

It is evident that Graicunas' conclusion is extremely speculative. It scarcely considers the realities of any given situation such as the nature, extent and importance of such relationships (particularly in view of the Principle of Specialisation referred to above). It is an extremely interesting and original attempt at analysing a complex problem, but its validity must remain in doubt. It is perhaps remarkable that it should have survived so long in organisational theory — perhaps because it is one of the few definite statements that have been made. It is a statement that has not been widely transferred from theory into practice.

However, to return to Urwick's principles, the remaining two are the Principle of Balance and the Principle of Continuity. These are stated in such abstract terms that it is difficult to know how they should be interpreted in specific circumstances. Indeed, the Principle of Balance can be construed as a restatement of the Principle of Co-ordination in a different guise.

Furthermore, the Principle of Continuity could almost be taken as a statement of failure, in that it might imply that organising along the lines suggested by the principles would lead to the need to reorganise! However, a probably more accurate interpretation is that Urwick recognised that organisational tasks and objectives are subject to change, and this can result in a need for reorganisation to maintain efficiency and effectiveness.

It is perhaps worthy of comment that, if Urwick's principles are difficult to apply when considered in isolation from each other, the problems are increased when they are viewed as a set. Some appear to be in conflict one with another. Thus The Principle of Specialisation (taken in the sense in which Urwick appears to intend it) is at odds with The Principle of Co-ordination; the further specialisation is carried, the greater the need for effective co-ordination, which in turn implies more "generalists". Clearly, if Urwick's principles are to be accepted, there is a need for a balance to be struck between these requirements (a point which Fayol (*op. cit.*) appreciated), but there is nowhere any indications of how this balance can be found.

The root of this dilemma, the balance between functionalism and generalism can be traced back to Plato, particularly to *The Republic,* Book II, where Socrates says, "Consequently, more things of each kind are produced, and better, and easier, when one man works at one thing, which suits his nature, and at the proper time, and leaves the others alone". This, incidentally, is an excellent statement of the underlying philosophy of functionalism). The trouble with it, a trouble which still has repercussions today, is that it is an inadequate statement about the nature of people. It is not true, by and large, that a person's nature (to adopt Plato's term) is such as to suit him for one activity only; most people are equipped to be more than adequately competent in a variety of different fields. Indeed, some people have achieved outstanding results in what are normally regarded as quite separate areas. Instances which come readily to mind are Charles Dodgson (mathematician and children's writer), Samuel Johnson (lexicographer and wit), C. S. Lewis (theologian and novelist), Jackie Stewart (driving and clay pigeon shooting), Winston Churchill (politician and historian), Chris Chataway (athlete and politician), Josiah Wedgwood (businessman and scientist), and there are many others. Perhaps the most outstanding example is Leonardo da Vinci. However, at a more mundane level, people come equipped with an array of more or less developed talents, not a single functional skill. Any organisation which neglects this, as the functionalist school does, can at best hope to utilise only a fraction of the human resources at its disposal; at worst, it can expect its members to be frustrated and less than fully committed to organisational objectives. The functionalist view is founded upon an extremely limited view of human abilities and can therefore not hope to be fully successful. Perhaps the enduring attraction of the functionalist view is that it gives rise to tractable and readily manipulable organisations. Whether it is the best view for achieving organisational objectives is open to doubt.

To return to the theme of Urwick's principles, the logical consequence of The Principle of Responsibility and The Principle of Correspondence should not be allowed to pass without comment. It is the most elementary exercise in logic to deduce from those two that every superior has absolute authority over his subordinates. That this is an unacceptable state of affairs is demonstrated by many examples in history, most specifically perhaps by Magna Carta, and more recently in the rise of Trade Unions. It is unfortunate, therefore, that it should be encapsulated in what is still to a great degree the fount of modern managerial thinking.

It is not the contention here that Urwick believed in absolute authority — it is clear in context that he accepted limits on organisational authority, though these are not spelt out precisely. The point is that, taken out of context, as one should be able to do with fundamental principles, his statements lead to an unacceptable conclusion.

Much further work has been reported in this field, for example Blau and Scott (1963) Littever (1963), Edwards and Townsend (1961), Miller and Rice (1967), amongst many others. It is not possible to review all the literature in depth here, but the general overall content of the majority is further exploration and refinement around the principles expounded by Urwick. Amongst the more interesting contributions has been that of Brown (1971), who amongst other issues, introduces the concept of more than one structure of roles being required within an organisation, for different purposes. He identifies in particular operational systems, representative systems and legislative systems. He also makes a very careful analysis of role structures and role relationships, laying great emphasis on accurate role descriptions.

Another development of interest has been the realisation that organisation structure is interdependent with information flow networks. Since it can be argued (though perhaps not entirely successfully) that this has arisen from the influence of cybernetic concepts, discussion of this development will be postponed.

Thus, the overall managerial view of organisation structure is one still based on specialisation, either function or process-oriented (though there are some experiments with project-oriented organisations), and that the organisation chart is an adequate tool for its design. Whether acknowledged or not, Urwick's work still exerts a major influence in this field.

In view of the difficulties with his approach outlined above, it is encouraging to find that some of the problems are being acknowledged. Thus, for example, Newman (1973) writes:

Furthermore, I think that the stage has been reached in some situations where the organisation will have to be changed, away from what is desirable in purely organisational terms, in order to enable real human managers, with their fallibilities, their limitations, to be relatively competent, relatively effective in their work.

A more comprehensive condemnation of current organisational theory and practice is difficult to find. Nor is it an isolated view. Duerr (1971) writes:

The need to escape from the hierarchy straightjacket is getting more and more common in business (just as it is in the army) as time goes by, with the introduction of more and more staff jobs, themselves made necessary by the advancing complexity of modern corporations . . .

It is perhaps significant that Newman and Duerr represent two quite separate schools in the study of organisation, what might be termed the "academic" and the "practical" view respectively. When two such disparate views emerge with the same general conclusion, it is fairly sure indication that the conclusion reached deserves serious consideration.

Perhaps the only general view that emerges from the study of organisation structure is that the structure needs to be adapted to the particular needs and circumstances of the individual organisation. Unfortunately, there appears to have been no attempt to be specific about what circumstances imply the need for certain types of organisation. (Woodward's study (*op.cit.*) comes closest to doing this, but it was very restricted in its range, and, as mentioned above, had very little in the way of yardsticks for effectiveness of organisation.)

In view of this lack, it is worthwhile to attempt to categorise at least some of the variables that might reasonably be expected to have a significant role in determining the type of structure appropriate to a given organisation. Such an attempt does not necessarily imply acceptance of the view that optimum structure is specific to local circumstances, but it is a necessary step in examining the truth of the proposition.

There would seem to be at least five major variables that could be used in classifying organisations. These are (a) the degree to which it is self-financed, (b) the degree to which it is "authoritarian", (c) the degree to which its sub-units communicate, (d) the degree to which its operations are continuous, and (e) the degree to which the environment is stable.

The degree of self-financing appears to be of importance, if only because it encapsulates a distinction that is generally held to be important, the distinction between business and non-business activity. Virtually every organisation needs finance to support its activities; it can obtain this either by the sale of goods and services (business activity) or by grant of funds from some external body. This would seem to be a distinction of degree, not of kind. Businesses obtain funds from external sources (bank loans, government grants, etc.) as well as from profit from operations. Equally, grant-aided organisations may derive some income from their activities (Arts Councils, and nationalised industries, for example). The principle difference that this would seem to make to an organisation is the extent to which it can allow the organisation to make its own autonomous decisions without reference to an outside authority. Thus, one significant role of profit in a private enterprise is to allow it to continue to determine its own future course of action. Exactly what influence, if any, this will have on its organisation structure is difficult to say without further investigation, but until evidence to the contrary is available, it would be as well to include it as a parameter.

The second proposed variable, the degree to which an organisation is authoritarian, requires some explanation of the term used. "Authoritarian" is not used in its generally accepted sense, but no reasonable alternative seems available which is not subject to equal confusion. The basic distinction which it is intended to convey is between the type of organisation which has been set up to serve the purposes of one individual (or a small group of individuals) where authority basically resides at the top of the hierarchy, and a different type of organisation set up by a large number of individuals in order to further some common purpose, where the authority basically resides at the base of the hierarchy. This latter type is typified by Trade Unions, though it is a category that in principle includes all forms of democratic representative bodies, including the House of Commons (but not, interestingly enough, the House of

Lords). Once again, in practice this is a distinction in degree rather than in kind, it is rare to find an organisation that is purely "authoritarian" or purely "democratic". Additionally, although the extremes of the dimension represent quite different needs, it is again not immediately apparent that they require different structures. This is perhaps in part due to the fact that basically there is only one model of structure available, that of hierarchy.

The third proposed variable, the degree to which the sub-units communicate, seems more immediately relevant to organisation structure. That there are differences in communication between operating units seems reasonably clear. For example, in naval operations, it may well be the case that two vessels will not interchange any communication, although both are carrying out the same mission. On the other hand, in a business operation, the sales force and the production process may be in virtually continuous communication (though it is tempting to be facetious and remark that there may well be occasions when it is open to doubt whether sales and production are attempting anything in common). Clearly, the communication needs in such disparate circumstances are quite distinct, both between co-operating sub-systems and as regards reporting procedures to higher levels of control and command. If organisation structure and communication needs have any bearing on one another, then it is logical to conclude that differences in structure are to be expected, and may well be justified.

The degree to which operations are continuous (the fourth proposed variable) does not seem to have gained much mention in the literature. The paradigm seems to be taken as the mass-production industry, where it is important to keep activities going continuously. There are, however, many organisations for which this is a misleading parallel. The prime example is that of the armed forces who (it is to be hoped) are employed in their primary task of fighting at only rare intervals, and other duties that they carry out are basically "filling in time". There are, however, many other organisations that function basically on an intermittent basis, such as a football club, the Fire Service, fish canneries, frozen vegetable suppliers, and so on. Many businesses are markedly seasonal (toys, publishing, Christmas cards, etc.) and are closer to an intermittent than a continuous operation. It is reasonable to suspect that the organisational requirements for the two extremes may be different; certainly the problems will be different.

The fifth and final proposed variable is the degree to which the environment is stable. It must be remembered here that the "environment" is a function of the organisation; each organisation finds itself in its own environment, and it is the stability of this that is important. Even two firms in nominally the same business may find themselves in markedly different environments — for example, the circumstances attached to British Leyland are quite different to those surrounding Rolls-Royce. The latter has secured an exclusive niche in the market, relatively stable demand, with little direct competition. The same is not true of British Leyland, (though at one time it may have been). The stability or otherwise of an environment could be expected to have consequences for organisation structure. At one extreme, with a rapidly changing and unpredictable environment (the two conditions are not tautologous) the emphasis should be on rapid response. Here again the military situation is the paradigm. In a stable environment, the emphasis needs to shift to considered action and the long-term view, and the paradigm is perhaps the Civil Service or a basic industry such as coal or steel. The demands for information processing and decision procedures at these two extremes are distinct, and may well be reflected in organisation structure.

There is one omission from this list of variables that may cause some surprise, the variable of size of organisation. This has been omitted because it does not appear to be of such fundamental importance as the issues that have been raised. The basic form of organisation, it can be argued, should be derived from considerations such as those listed above; the size of an organisation may well require the basic form to be replicated at different levels, and greater specialisations within this form. Nevertheless, the basic structure is not a function of size.

There are no known studies of organisation using the variables set out above, and further research is needed to investigate their usefulness as classifications of organisation types. There are two major difficulties in the way of such research. Firstly, there is only one major model of organisation available, that of hierarchy, and it may be that different concepts are needed. Secondly, organisations change in the course of time; reorganisation is a frequent concern of senior managers.

In conclusion, it can be seen that the basic form of organisation today is that of a hierarchy, involving the concepts of line of command, authority, responsibility, and delegation, and this can be traced back to the original military tradition. Modern thinking is beginning to question some of these ideas, but any cybernetic description of management must be capable of including the concept of hierarchy, as well as possible alternative forms. It is, of course, desirable that a cybernetic model will account for the phenomenon of hierarchy in more fundamental terms, as well as point the way to other structures.

The position is well summarised by Woodward (*op. cit.*):

The danger lies in the tendency to teach the principles of administration as though they were scientific laws, when they are really little more than administrative expedients found to work well in certain circumstances but never tested in any systematic way.

2.3 The Dynamics of Organisations

Given that organisations are hierarchical structures of people and equipment, what do they actually do? What are their activities, what roles do people play in them? These are questions of dynamic function rather than static structure, although the two aspects are closely related.

Clearly, each individual organisation is unique in this respect if considered at a sufficiently detailed level. However, it has been found that there are sufficient similarities between organisation to enable useful classifications of activities to be made.

The pioneer in this field was again Fayol (*op. cit.*). He produced the following list of activities, which he suggested were present in all industrial undertakings:

1) Technical activities (production, manufacture, adaptation)

2) Commercial activities (buying, selling, exchanging)

3) Financial activities (search for, and optimum use of, capital)

4) Security activities (protection of property and persons)

5) Accounting activities (stocktaking, balance sheet, costs, statistics)

6) Managerial activities (planning, organisation, command, co-ordination, control)

It must be remembered that Fayol was referring to industrial concerns; there are organisations which do not undertake all these activities, but specialise in one or two of them — retailers and finance houses, for example.

A point of special interest is that Fayol specifically includes management activity as a distinct classification. It is worth remembering that his original work was published in 1916, based on experience gained during the late 1800's, when industry was only just beginning to move out of the era of the individual entrepreneur into the era of the corporate enterprise. It says much for Fayol's acute perception that his concepts have withstood the passage of time and still remain valid today.

Furthermore, he was not content to identify management as an activity and leave it at that. He spelt out quite specifically what he saw as the functions of management. Those were:

1) To forecast and to plan, which means to examine the future and draw up plans of action.

2) To organise, which means to build up the structure, both material and human, of the undertaking.

3) To command, which means to maintain activity among the personnel.

4) To co-ordinate, which means to bind together, unify, and harmonise all activity and effort.

5) To control, which means to see that everything happens in conformity with established rule and expressed command.

This analysis still remains as the basis of modern thinking on management theory. It has been extended and modified, but never seriously challenged. It is remarkable that Fayol's analysis of management, which is the first known attempt at a theory of management, has survived largely unscathed. He even provides a definition of good management — "to get the optimum return from all employees of his unit in the interest of the whole concern" — which is still relevant today.

Many other writers have contributed observations on the functioning of organisations. The divergence of views available is difficult to summarise adequately, but some of the main themes can be seen in the works of Barnard (1948 i and ii), Brown (1960, 1962, 1971) and Bakke (1950, 1953, 1959).

Barnard's view of the nature of organisation ("a system of consciously co-ordinated activities or forces of two or more persons") has already been mentioned. His view of the functioning of organisations centres around the concepts of purpose, communication, and commitment, which can be related to Fayol's notions of planning, co-ordinating and commanding. His view of purpose is interesting in that he sees the purpose of an organisation not in abstract terms such as "survival" or "profit" but as the production of a specific item or service, and as such is an extremely pragmatic approach.

Given that co-ordination of activities is required, it follows, Barnard argues, that acts of communication are necessary so that purpose can be translated into action. He views communication in a very broad sense, not restricted to verbal or written media, and this leads him on to consider the "informal organisation", — the network of communication that supplements the manifest organisational structure. Furthermore, Barnard recognises that the degree to which an individual will accept the organisational purpose will vary from person to person and from time to time, and he sees an important part of the functioning of an organisation is the securing of sufficient commitment to its purposes from its personnel. He takes a somewhat pessimistic view of the nature of this process, and seems to feel that this commitment is hard to obtain in modern societies.

Basically, Barnard seems to expand on Fayol's principles, and introduce some of the complexities of these principles in practice. In particular, he emphasises that psychological and sociological forces have their part to play in the functioning of organisations, and acts as a precursor to more modern investigations in industrial psychology and the like.

Brown, writing on the basis of his own experience of management, takes an interesting and individual approach, complementary to Fayol rather than directly derived from him. He sees the functioning of an organisation in terms of social systems, of which he identifies three, namely:

1) The Executive System
2) The Representative System
3) The Legislative System

The Executive System is meant to comprehend the structure of roles usually referred to as the organisation chart or hierarchy, and Brown maintains that this exists irrespective of people; people may come and go, but the roles do not disappear. (Interestingly, here he is at variance with Barnard, who is prepared to concede that organisation may be tailored to the individuals available.) He also points out that the design of this Executive System may have conflict built into it, and that friction between individuals can arise because of this, friction which is all too easy to put down to "a clash of personalities". He does not seem to agree that such conflict can be beneficial to an organisation in the long term, by providing a source for change and development.

At root, Brown's concept of the Executive System is in accord with the views of Fayol and Weber, that of an ordered hierarchy, but he examines it in considerable detail, introducing variables of Personnel aspects (organisation and personnel) a Technical aspect (production techniques) and a Programming aspect (balancing, timing and quantification of operations). He is particularly concerned with the role of specialists in these aspects and their relation to the actual work process, and elaborates on what he feels to be suitable structure to accommodate these needs.

However, rather more interesting is his identification of the Representative system which, he maintains, will always exist alongside an Executive system and acts to convey the feelings of subordinates upwards to superiors, in contrast to normal channels which convey information from superior to subordinate. Brown points out that this system may not be explicitly recognised, but he maintains that it always exists. Commonly, these days, such a structure is given formal recognition (as a Staff Council or some such body, or a Trade Union). The point of interest, however, is the contention that such a system is an integral part of the functioning of any organisation, and (although Brown does not directly say so) is quite distinct from Barnard's informal organisation.

On top of this complication of the view of organisation, Brown adds a further system, the Legislative system. This he envisages as an interaction between shareholders, directors, customers, the Executive system and the Representative system. He maintains that the joint power of these systems, and the interaction between them puts limits on what a company can do — in effect, legislates for the company, and hence the title given to this system.

It can be argued that this last analysis is not wholly convincing. For instance, Brown's other two systems consist of a set of structural roles, whereas his Legislative system is a process of interaction, and is thus different in kind. At a more mundane level, it is rare for shareholders to exert any direct influence over the actions of a company, and virtually impossible for customers to do so. Nevertheless, a company does need to bear in mind the attitudes of shareholders and customers alike, even if it does not negotiate directly with them. It is certainly a valid point that organisations do not exist in a vacuum, and are subject to powerful influences from outside which severely circumscribe its freedom of action. It is not necessary to limit these influences to just simply shareholders and customers; government and competitors, for example, play just as significant a role.

A different approach to the functioning of organisations is typified in the work of Bakke. His is a somewhat more academic approach, and his aim is more to provide a theoretical framework of analysis, applicable to all types of organisation, not necessarily just business and commercial activities. He approaches this task by considering the basic resources which any organisation needs, a rather different line of attack from many other analysts. These he identifies as:

Human Resources

Material Resources (including plant and equipment)

Financial Resources

Natural Resources (i.e. not processed by human activity)

Ideational Resources (including the language used to communicate these ideas)

It is possible to question whether all these resources are essential to every organisation (for example, does a bank need natural resources, does a Ramblers' Association need financial resources?) but these are rather forced examples. Of more interest is the inclusion of Ideational Resources. Where these originate if not from the human resources is not clear, but the main aspect of interest is the implicit acknowledgement of the importance of information processing to the functioning of an organisation. This is a distinctly different thread, not found in many other schemes of analysis, yet its importance should not be allowed to pass unmentioned.

Bakke introduces the concept of the Operational Field of an organisation, which can be considered closely analogous to what many others term "the market", which he also appears to consider to be a resource of the organisation.

He then goes on to consider that the functioning of an organisation can be regarded as the operation of Activities on these resources and further that these activities can be classified under five headings, namely:

1) Perpetuation
2) Workflow
3) Control
4) Identification
5) Homeostasis

Perpetuation activities are those acts designed to ensure that the organisation continues to have access to the necessary resources. Examples include recruitment of new personnel, or the issuing of more shares. Workflow activities include all those acts which are necessary to create and distribute the output of an organisation, be it goods or services. Examples include assembly operations, driving vehicles, and sales activities. Control activities are specified as designed to co-ordinate and unify, and are further subdivided into:

1) Directive activities which initiate action, such as deciding what work will be done and to what standards.

2) Motivation activities, rewarding or penalising behaviour.

3) Evaluation activities, such as reviewing and appraising performance, or comparing alternative courses of action.

4) Communication activities, providing people with the premises and data needed to perform the job.

Identification activities are what might be termed image-building, presenting an image of the organisation both to its members and the environment, with the aim of promoting the character, or "charter" as Bakke terms it, of the organisation.

Homeostatic activities are those which are designed to preserve the dynamic equilibrium of the organisation, arranging and regulating the other four types of activity so that the organisation is maintained in existence. Again, further subdivisions of this type of activity are introduced as follows:

1) the Fusion process
2) the Problem-solving process
3) the Leadership process
4) the Legitimisation process

In postulating a Fusion process, Bakke accepts Barnard's premise that there will be conflict between the aims of individuals and the aims of an organisation. The Fusion process is the name he gives to the way in which these differences are reconciled, enabling people to co-operate. He takes this concept further and applies it to the relationships between the organisation and other outside bodies. Rather than a series of specific acts, Bakke seems to regard this Fusion process as a useful framework for categorising and understanding some otherwise inexplicable activities.

The Problem-solving process is the term applied to the continual solving of non-routine problems, and an attempt is made to provide a sequence of steps used in logical problem-solving. This is a particularly interesting aspect of Bakke's analysis, in that it recognises problem-solving as an activity that occurs within organisation as a necessary part of their activities.

Finally, the Legitimisation process aims to justify and get accepted both the purposes of the organisation and the means adopted to pursue them. This can range from the registering of Articles of Association at one end of the scale to Alfred Sloane's reported dictum of "What's good for General Motors is good for the USA", at the other. It is an expression of the idea that ultimately an organisation cannot survive without acceptance by society at large.

It is evident that there is some overlap in Bakke's categorisation — for example the precise boundary between Perpetuation and Legitimisation is not altogether clear, nor are the boundaries between Control and Homeostasis precisely defined. Nevertheless, the concepts do provide a framework for surveying the functioning of organisations.

The foregoing authors are not an exhaustive list of people who have contributed to the study of organisations, but it can be maintained that they are reasonably representative of the main strands of thought. Taken as a whole, it can be seen that the basis was laid by Fayol, and others have followed his lead. Most of the concepts and categories introduced by other workers can be related to Fayol's, with rearrangements to suit the differing points of view of other writers, combined with elaborations and further elucidations on particular points. Barnard, for example, contributes the concept of purpose, and conflict of purpose, together with the notion of the informal organisation. Brown elaborates on the variety of role-systems and structures within an organisation, and the importance of psychological and social systems within organisations, as well as introducing the concept of an organisation being regulated at least in part by its environment. Bakke elaborates to some extent on this relationship between organisation and environment, in particular bringing out the point that organisations attempt to influence the environment as well as vice versa. His other major contribution, in the present context, is the introduction to the idea of information processing and problem-solving as an essential part of organisational activity.

Viewed as a basis for a theory of organisation, these works would appear to suffer from a serious limitation. They are all based on reported experience, and represent attempts to classify that experience into general categories.

What is lacking are any underlying concepts at a more atomic level of detail that would in the first place suggest a more fundamental scheme of classification and in the second place enable a testable model to be constructed.

Nevertheless, taken together, these writers present a useful picture of the functioning of organisation, and of some of the complexities that need to be accounted for in a theory of organisation.

2.4 The Control of Organisations

In the previous section, "management" was mentioned as one of the functions of organisations, and its tasks set out under broad headings. It is clear that, if there is a controlling function within an operation, then it resides with this group of individuals. Thus, management is a central concern, and many people have written on the topic either to report on the reality of managerial life or to offer more or less comprehensive theories of management.

One of the early pioneers in this field was Taylor (1903, 1911, 1947), who founded the movement known as Scientific Management which was an attempt to subject the process of management to the scrutiny of objective scientific study. Apparently, his stimulus for doing this was his observations of inefficient production and of antagonism between workers and management. This was completely at odds with his conception of an organisation as a co-operative enterprise. For him, there was no conflict between high wages and high profits. As he wrote (1911), "The principal object of management should be to secure the maximum prosperity for the employer, coupled with the maximum prosperity of each employee." This, of course, in today's terminology, implies high productivity.

He identified three obstacles to this goal:

1) Belief by workers that any increase in output would lead to unemployment, a belief which Taylor thought fallacious.

2) Defective systems of management, which made it necessary for workers to restrict their output to protect their own interests.

3) Inefficient, rule-of-thumb, effort-wasting methods of work.

To overcome these, Taylor proposed use of "Scientific Management", by which he meant firstly a systematic study of work to discover the most efficient way of performing a job, and then a systematic study of management, to discover the most efficient methods of controlling the workers.

To achieve this, Taylor proposed his four underlying principles of management, which were:

1) The development of a true science of work. This revolved around establishing "a fair day's work", acceptable to both workers and management, and for which the worker would be highly paid. This high pay, made possible by high productivity, was an essential element in Taylor's thinking, the due reward for accepting scientific management.

2) The scientific selection and progressive development of the workman. In order to ensure that the worker could achieve high output, Taylor believed that it was first of all necessary to select people with the physical and mental qualities required by the job, and then to train them systematically to become "a first-class man". It is of interest that Taylor thought that this training should be a continuous process, to develop the worker to the highest level of which he was capable.

3) To bring together the science of work and the scientifically selected and trained men. This Taylor saw as a revolutionary change of attitude, particularly for management. He found little resistance among workers to learning to do a good job for good pay.

4) The constant and intimate co-operation of management and workers. Taylor's concept here was that management took over all the work for which they were fitted than the workers. (There is an interesting parallel here with the views of Plato (*op. cit.*) on the organisation of the city state.) The tasks which he had in mind were those such as specification and verification of methods, and quality, and continuous control of the worker. He maintained that with this close personal contact, opportunities for conflict would be almost eliminated, since the operation of authority would not be arbitrary. The manager would be continually showing that his decisions were subject to the same discipline as the workforce, i.e. the scientific study of work.

Taylor's thinking was developed by a number of people, notably Gantt, Gilbreth, and Bedaux, and led eventually to the group of techniques known as Work Study and/or Industrial Engineering. It can hardly be claimed that they have done justice to his ideas. They have concentrated almost exclusively upon one limited aspect of his work, that of establishing norms for output using improved methods, and almost totally ignored his other principles. Taylor's own contribution was much broader than this, and many of his ideas are still extremely relevant today. He still stands as the pioneer of the application of the scientific spirit of enquiry to the problems of management.

A different approach to management, or perhaps an examination of a different aspect of the subject, is exemplified in the work of Follett (1920, 1924; collected papers 1941, edited by Metcalf and Urwick). Her approach was centred much more on the human interactions within organisations, and especially the attempt to analyse the fundamental motives involved in human relationships. Her aim in this was to answer two questions:

1) What do you want men to do?

2) How do you scientifically guide and control people's conduct in work and social relations?

This work led her to an appreciation of the value of psychology, then a new discipline, and she was a pioneer in applying this tool to the analysis of organisational and managerial problems. The central problems for her were those arising from the need to reconcile individuals and social groups, and to weld these groups together into a cohesive whole. She too formulated four principles:

1) *Co-ordination by direct contact.* Follett maintained that the responsible people must be in direct contact, regardless of their position in the organisation. This she applied to horizontal communication across a hierarchy as well as vertical communication.

2) *Co-ordination in the early stages.* In order to increase motivation and morale, people who will be affected by decisions should be brought into the decision-making process at an early stage — before decisions are formulated, not afterwards.

3) *Co-ordination was the "reciprocal relating" of all factors in a situation.* All factors have to be related to one another, and these inter-relationships must themselves be taken into account.

4) *Co-ordination as a continuing process.* The making of management decisions is a continuing process, not a series of isolated events. Many individuals contribute to the making of a decision, and the concept of final responsibility is an illusion. Authority and responsibility should derive from the actual function to be performed rather than from position in an hierarchy.

As can be seen, Follett's main concern was with the integrative aspect of management, with arranging a situation so that people co-operate of their own accord. She laid great stress on her concept of "The law of the situation"; she maintained that conflict could be avoided by the joint study of facts, from which the law of the situation would emerge. This in turn would lead to an agreed course of action.

It is possible to criticise Follett's views as being largely restricted to one aspect of management, and based on a somewhat idealistic view of human nature. Nevertheless, her contribution of the concept of partnership, the joint rational approach to problems, brought a new element into thinking about the management process. In particular, her attention to the importance of psychology initiated a major thread in the understanding of organisations.

A complete contrast to Follett's approach can be found in the work of

Simon (1958, 1960a, 1960b). To Simon, the complete essence of management lies in the taking of decisions, and he has devoted a great deal of attention to the way in which decisions are taken, and the effectiveness of these processes. In outline, he identifies three main stages in reaching a decision:

1) Finding a problem that requires a decision — an investigative activity.

2) Inventing, developing and analysing possible courses of action — a design activity.

3) Selecting a particular course of action from those available — a choice activity.

In practice, the process may be much more complex than this, involving iterative loops and many levels of analysis, but the same three stages can still be discerned. Likewise, the implementing of a decision that has been made can be regarded as a further set of problems and decisions.

Over and above this, Simon is concerned to attack the view that managerial decisions were taken on the basis of arriving at a rational evaluation of the maximisation of economic return. To allow for the element of emotional and unconscious factors in human decisions, he introduced the concept of "satisficing" — of a decision being "good enough". This allows a gross simplification of the decision-making process, and reduces the number of factors that have to be considered.

He furthermore distinguishes two types of decision lying at the ends of a continuum. These are programmed and non-programmed decisions. Programmed|decisions are routine and repetitive, and frequently there is a definite procedure for dealing with them (an algorithm). On the other hand non-programmed decisions are new and unstructured, with no definite method to resolve them, (heuristic decisions). He foresees that modern developments in mathematics and computing will make it possible for an ever-increasing proportion of unprogrammed decisions to be made on computers, until eventually all aspects of organisation will be automated.

It is possible to disagree with this conclusion on a number of grounds including the difficulties encountered in heuristic programming, and the probable psychological reaction against a computer running a business. Nevertheless, Simon provides important insights into the executive decision process.

It is impossible to discuss the management of organisations adequately without mentioning the work of Urwick and Brech (1947, 1950, 1957, 1963). Their contribution was not so much any specific innovation as a collation of the work of many others, covering an extremely wide range of topics, and bringing a degree of coherence to the subject. Additionally, they were extremely active in promoting the practical application of the growing body of management theory. Their work has surveyed the field of management much more comprehensively than has been possible here and has provided the foundation for much further work.

2.5 People in Organisations

It has been mentioned several times already that people are an essential element of organisations. An important area of study has been the way that people actually behave within organisations, and the consequences of this behaviour.

This area of interest can be traced to the work of Mayo (1933, 1949), who carried out the well-known Hawthorne studies. In these experiments, a series of changes were introduced into a work situation, and the effect on output noted. The result was that output was increased, but this could not be attributed to the changes — for example, one "change" was to revert to the original, pre-experiment, conditions, and this also resulted in increased output. Eventually (though not originally) Mayo came to the conclusion that the rise in output resulted from a change of attitude amongst employees, a change brought about by their participation and involvement in the experiments. Coupled with other investigations on attitudes, motivation, and morale, this led to the concept of the informal working group (a different notion to the informal organisation), and a recognition that the group exerted considerable pressure on individuals within it to conform to expected behaviour. Mayo also identified different logics in the attitude of workers and management; the former was a logic of sentiment, the latter a logic of cost and efficiency. In such a situation, it is all too easy for conflict to arise.

Mayo devoted much time to trying to find ways in which this conflict could be resolved. Although he was unsuccessful in this aim, the true measure of his success is in founding what can be termed the Human Relations school of thought, and the use of the social sciences as a tool of investigation in organisations. He threw great light on the influence of the "human factor" in the work situation.

A different aspect of problems of people and organisation can be found in the work of Jaques (1951, 1956, 1961). He worked at Glacier Metals, and much of it was in association with Brown (*op. cit.*) on the topic of organisation structure. However, his distinctive contribution was in his approach to the analysis of work and responsibility. He divided work into two elements, a "prescribed content" and a "discretionary content". The prescribed content was exactly specified, leaving no need for judgement on the part of the worker. The discretionary content was more loosely specified, and required a degree of judgement from the worker. It was Jaques' contention that all jobs had some element of discretionary content, but the proportion of this varied widely from job to job. Furthermore, the discretionary content varies in the length of time that needs to elapse before the effectiveness of an act of judgement can be effectively reviewed. From this, Jaques developed the concept of "time-span of discretion" and the use of this concept to evaluate the importance of a job. He found that time-span of discretion increased as level in the organisational hierarchy increased. In later work, he applied these results to the calculation of wages and salaries, and particularly to the problem of equitable differentials in pay at different levels of the organisational hierarchy.

Although his work has received little follow-up — perhaps because it was seen as just another payment scheme — it deserves attention as a pioneering effort in the application of science to management. It is an attempt to produce a rational basis for the quantification of managerial work.

Mention also needs to be made of Argyris (1957, 1960, 1962). He examined the role of an individual in an organisation in terms of the conflict between the needs of the two. He maintained that such a conflict was unavoidable, and the result was mutual adaptation, together with the development of informal groups.

The conflict he saw was rooted in the development of an individual from infancy to adulthood, maturity and independence. A mature individual will strive to set his own goals, and will allow others to do the same. Additionally, having set his goals, he will strive to achieve them — and in doing so, will adapt to his environment — a process Argyris terms "self-actualisation". Against this, the basic characteristic of a formal organisation is rationality. Ends and means are expressly given, goals and activities are imposed. The results of this for the individual are that his job requires only a few, shallow, abilities, he becomes dependent upon his leader (i.e. passive and subordinate), his time-perspective is shortened, and, perhaps most important, his goals are defined and controlled for him. Together, these create the conditions for psychological failure.

To adapt to this situation, the individual can adopt one of four courses. He can leave the organisation; he can rise in the organisation, he can use psychological defence mechanisms; or he can become apathetic and disinterested. These adaptive responses are re-inforced by informal groups. Commonly, the observable result is lack of interest and restriction of output. This in turn can set up a vicious circle as management becomes more autocratic and authoritarian.

Argyris suggests some possible means of alleviating this conflict. These include "job enlargement", allowing the worker to use more of his abilities, a more democratic approach by management, and particularly a more skilled and sensitive approach to human relations by managers. (It is interesting to note the parallels between these views and those of Taylor (op. cit.). To this end, he proposed special training for managers in human relations.)

In his own way, Argyris has made an important contribution to the understanding of the interaction between people and organisations. His work is separated from that of Mayo by his comparative emphasis on the psychology of the individual, as contrasted to Mayo's emphasis on the importance of group processes.

A feature of the views of people in organisations put forward by Mayo and Argyris is that it basically relates to the viewpoint of a subordinate, particularly of a worker. "Management" appears in their works as a nebulous, and somewhat forbidding, entity, almost a *deus ex machina*. There is little acknowledgement of the fact that "management" consists of people too, and little effort to examine the psychological factors that drive the behaviour of managers and executives. Attention to this aspect of organisation can be found in the writing of McGregor (1960) and Likert (1961).

The basis of McGregor's work was an examination of the underlying assumptions about human behaviour that appear to govern managerial behaviour, particularly the type of managerial behaviour prescribed by traditional management theory as expounded by Fayol, Brech, and others referred to above. He summarised these assumptions, under the heading of "Theory X", as follows:

1) The average human being has an inherent dislike of work and will avoid it if he can.

2) Because of this human characteristic of dislike of work, most people must be coerced, controlled, directed, threatened with punishment to get them to put forth adequate effort toward the achievement of organisational objectives.

3) The average human being prefers to be directed, wishes to avoid responsibility, has relatively little ambition, wants security above all.

Theory X has persisted for a long while — indeed, the work of Mayo suggests that it is a self-fulfilling prophecy, in that organisations based on Theory X will produce behaviour in line with its assumptions. However, McGregor felt that Theory X was not necessarily true, a view supported by observation. He proposed an alternative view, which he called Theory Y, in which the basic assumptions about human behaviour were:

1) The expenditure of physical and mental effort in work is as natural as play or rest.

2) A person will exercise self-direction and self-control in the service of objectives to which he is committed.

3) The most significant reward that can be offered to obtain commitments is the satisfaction of the individuals self-actualising needs. This can be a direct

product of effort directed towards organisational objectives.

4) The average human being learns, under proper conditions, not only to accept but to seek responsibility.

5) Many more people are able to contribute creatively to the solution of organisational problems than do so.

6) At present, the potentialities of the average person are not being fully used.

He went on to examine how the adoption of this theory would affect the running of organisations, particularly in such areas as performance appraisal, salaries, promotions and the like. Not surprisingly, since Theory X and Theory Y are diametrically opposed, he found that many changes could be called for, which goes some way to explaining why his views have not been widely implemented, although lip-service is often paid to them.

A very similar view was put forward by Likert, though in contrast to McGregor his work was based on research findings. These findings showed that low-efficiency groups tended to be in the charge of supervisors who were "job-centred", i.e. supervisors who concentrated on keeping their subordinates busily engaged in going through a specified work cycle in a specified way. (This is an attitude clearly derived from Taylor (*op. cit.*). Whilst there were some highly productive groups led in this style, they were exceptions, and were not without problems. Generally, the effective groups were supervised by managers who concentrated more on the human aspects of their subordinates problems, and on building effective working groups. They were more concerned with getting high targets accepted than with the details of the work. In particular, these supervisors were interested in their subordinates as individual people, rather than as work-producers.

A common theme in the work of both McGregor and Likert was the view that essential role of management was to provide the support and assistance required by individuals to enable them to function. Between them, they cast light on the psychological processes of managers, complementary to the work of Mayo, and Argyris on the psychology of employees. The work of Jaques forms a distinctively different thread, which to some extent forms a bridge between the others.

2.6 The Environment of the Organisation

An essential fact about organisations is that their functioning cannot be fully understood by regarding them in isolation. They exist in, interact with, and are a part of, a much wider culture and society.

Several writers have been concerned with this complex relationship, from differing points of view. The political aspect of this relationship is expressed in the work of Burnham (1941). Although not an original view (as he himself says) he analysed the relation between organisations (specifically, business organisations) in Marxist-capitalist terms. The conclusion he came to was not that capitalism was giving way to socialism but that a new class was emerging in society, the managerial class, who were in the process of becoming the dominant social group. Increasingly the wealth of society was being produced by organisations, and organisations were controlled by managers; the role of shareholders, financiers, and the boards of companies were becoming less and less influential, and as a result power and influence were being concentrated in the hands of managers. Increasingly management was taking on the trappings of power, and influencing the political and legal process.

Burnham saw this as a continuing trend, which would have important repercussions on society. There would be a move away from the individualistic ideology of capitalism towards the concept of the state, with increased emphasis on planning, security, duty and order, rather than freedom, jobs and individual rights. This trend he named "The Managerial Revolution". Although it cannot be claimed that all his predictions have been realised — perhaps because of the rise of Trade Union power — his analysis is an example of the powerful forces involved in the relation between organisations and society.

A somewhat similar approach to the relation between society and organisation can be seen in the early work of Drucker (1939, 1943, 1946). He took as the archetype of modern organisation the large corporation embodying a mass-production plant. He saw the central dilemma of such organisations as being that although economics was the driving force behind such institutions, economic activity for its own sake makes no sense; account must be taken of wider social, ethical and moral considerations, or the whole structure would wither and die.

To overcome this dilemma, Drucker maintained that a "Functioning Society" was required, which would involve three things. Firstly, the individual must have a definite social function, which would largely be defined in terms of his occupation. Secondly, he must have a recognised social status. Thirdly, and most importantly, these two must be shown to be accepted, by legitimising the distribution of social power.

He contended that, for power to be wielded legitimately, it must be justified in terms of the basic value structure of society, and further that this was no longer true in Western society as a whole. The original basis for managerial authority was derived from individual property rights, but with the rise of large corporations this was no longer valid. Managerial power, in practice, was not controlled or limited by shareholders, for various reasons. Thus management power was unfounded, unjustified, uncontrolled and irresponsible, since it was not based on a principle which was accepted by society as legitimate. Hence, management must be legitimised.

To achieve this, it was Drucker's view that organisations needed to pay heed to ethical factors as well as economic factors, and fulfil their social obligations in addition to pursuing profit. The key ethical considerations were, for him, equality of opportunity and individual dignity. The alternative to this type of solution was the disintegration of society as it existed, and its replacement by a totalitarian state.

Although this analysis parallels that of Burnham in many respects, the important element it brings in is the relevance of ethical and moral considerations to the running of a business. This theme is taken up and amplified by Whyte (1956), from the point of view of the individual. In a world which has become apocryphal, Whyte examines the conflict between the Protestant ethic of thrift, hard work, and independence, and the demands of the large organisation, which are expressed in what Whyte terms the Social Ethic. This Social Ethic emphasises the values of group identity, group belongingness and group achievement, together with a belief in science as a means of controlling human relationships. He examines at length the pressures upon the individual to conform to group behaviour, and the conflict between these values and the values necessary for attaining higher levels within the management hierarchy. It is Whyte's contention that such influence of the organisation over the individual is against the accepted moral ethic of society, and the individual must struggle to resist it.

In addition to influences such as these, the organisation, particularly a business organisation, must cope with external factors of economics. Economics is an area of study in its own rights, which it is not intended to pursue here. An introduction can be found in Tustin (1953) or Leontief (1941). The study of the economy is not directly germane to the issues to be discussed here, it is sufficient to identify it as a source of disturbance external to the organisation.

2.7 Summary

The foregoing has been intended as a survey of the main threads of what may be termed the received view of organisations, to identify their main characteristics. Organisations consist of a group of people who use resources to accomplish a common task (or set of tasks). These tasks can be regarded as consisting of several separate identifiable functions, which interact with each other, and within which people are assigned to specific roles. A function of particular interest here is that of management, whose role is broadly to plan, co-ordinate and control. (It is of interest to note that there is little attempt to justify the existence of management within organisations; it is more or less accepted, and its nature described.) Management also involves communication, problem-solving, decision-making and motivating. Particular problems arise within an organisation in reconciling the different interests of the people who constitute it.

These then are the general features of organisation. They suggest a complex system, involving equally complex goal-setting and control procedures, and as such merit serious cybernetic consideration. The managerial function is obviously of special cybernetic interest, and is the topic examined in the remainder of this book.

Before moving on, mention should be made of further work that has been done in the area discussed above. There has been a great deal published which it is not practical to discuss in detail. Much of it, however, develops the main themes set out above.

Firstly, there is much of what can be considered as reportage of management practice, usually admixed with some degree of didactic advice culled from experience, represented for example in the works of Stewart (1963), Townsend (1970), Parkinson (1958).

Secondly, the topic of organisation structure has been elaborated, by writers such as Newman (1968, 1973), Pfiffner (1960) and Barnes (1970).

Management techniques have received much attention. Amongst the major innovations can be counted the work of Humble (1970) in attempting to rationalise and structure objectives, the rationalisation of decision procedures via game theory and decision trees (see, for example, William (1966) or Kaufman (1968) or through applied logic as presented by Kepnor and Tregoe (1965); and the use of simulation, particularly in "management games". (See, for example, Eilou (1963).)

Similarly, problems of human relations in industry have been examined by workers such as Herzberg (1966) and Mazlow (1965, 1970). A development of particular interest has been the work of Blake (1969) in the analysis of managerial style and effectiveness.

3

The Cybernetic View of Organisations

3.1 A General Overview

The previous chapter discussed the general nature of organisations and set out the general classes of observable phenomena which should be accountable for within a cybernetic view of organisation. This chapter sets out to examine existing cybernetic approaches to the problem. In doing so, the decision has been taken to take a fairly broad definition of cybernetics, in order to confine discussion of a spectrum of approaches under one heading. In some cases, the dividing lines between a cybernetic view and a more traditional approach is somewhat hazy, and a matter of personal choice.

Three main themes can be discerned, the Operational Research approach, the General Systems Theory approach, and what may be termed for convenience the "pure" cybernetic approach — though again the dividing lines are hazy.

The Operational Research approach grew out of the success of applying scientific method to operational problems during World War II. Since then it has developed a philosophy of investigating situations through explicit modelling, usually using mathematical models, and manipulating the model to produce answers to specific problems. Several standard models have been developed to deal with common problems, such as stock control packages, linear programming techniques, network analysis, and queuing theory, as well as a large variety of more specialised models. An introduction to such work can be found in Duckworth (1962), Ackoff and Sansieni (1968) or Rivett (1968). A common feature of this area of study is that it is not so much concerned with problems of organisation as to provide decision procedures to solve particular problems facing particular managers at a particular time. A specially interesting study in this field is that of Ansoff (1965), who developed an analytical model for decision procedures at a very high level of management, dealing with problems of major investment in diversification of business.

The roots of General System Theory can be traced to von Bertalanffy (1956) and Sommerhoff (1950), working in the field of biology, who introduced the concept of the open system. At about the same time, Shannon and Weaver (1949) were developing information theory, a tool widely used in the analysis of systems. These concepts were soon applied to business organisations, in various ways. At one level, the general notion of a system as a complex interaction of functions and information flows was taken up by writers on business and applied at a descriptive level to the workings of business. A typical example can be found in the work of Hart (1964). A different approach can be found in what is usually termed systems engineering, as exemplified in Goode and Machol (1957) and Gagné (1962). Systems engineering is concerned with the detailed analysis (usually mathematical) of operational, on-line systems, and particularly with the initial design of such systems. Rarely, however, does it deal with matters of organisation and management. The closest approach to these problems is perhaps to be found in the work of Forrester (1961, 1968, 1969, 1971). He is concerned with the effects of time-lags on the dynamics system, the instabilities that can arise because of them, and strategies to reduce their worst effects.

A more managerially oriented application of the systems approach can be found in the work of Miller and Rice (1963, 1967), Emery and Trist (1960, 1965) and Cutcliffe and Strank (1968). These writers used a systems approach to various aspects of the managerial process, as distinct from concentrating on purely production operations.

As far as organisation and management are concerned, the distinctive contribution of cybernetics can be said to be the concepts of feedback and goals. These two ideas have found ready acceptance (though little critical evaluation) in managerial writing, to the point where it is rare to find a recent management text where they are not mentioned. The work of Humble (*op. cit.*) can be seen as a specific application of the concept of "goal" or "objective" in the organisational situation (whether or not it is a successful application is open to debate). Similarly, the work of Donald (1967) shows how these concepts are starting to be applied in the field of accountancy.

However, serious "pure" cybernetic attention to the nature and problems of organisations is comparatively rare. Even Wiener (1948) in his definition of cybernetics as "the science of communication and control in the animal and the machine" makes no reference to organisations, though his later book (1950) does make it clear that he was concerned about many of the problems that occur in organisations. Pask (1961) dismisses the whole subject in four pages, and appears to feel that all that is required is the application of a little elementary cybernetics to solve all problems. Thus he says (p. 110), "Cybernetics offers a scientific approach to the cussedness of organisations, suggests how their behaviours can be catalysed, and the mystique and rule of thumb banished", and proposes that management be replaced with an "evolutionary network" (i.e. a type of adaptive computer). He does acknowledge some of the possible problems, and concludes, "On this test, I shall accept the network if and only if it sometimes laughs outright, which, in conclusion, is not impossible." Unfortunately, he gives no specification for the network, nor does he discuss the problem of how the organisation will survive while the network is learning its job. Certainly, he does not appear to feel that there is any important distinction in principle between an organisation and a biological organism.

Ashby (1956, 1960) nowhere makes specific reference to organisations or management, though it is apparent that the concept of ultra-stability is of relevance.

Much of the published cybernetic work which refers to organisation is basically concerned with the application of principles to solve particular managerial problems (and is analogous in this sense to much OR work, as discussed above). Some examples of this can be found in Dewan (1969). Much of the work of Simon (1960, 1958) falls into this category, since he is concerned with the decision-making process, which is only one facet of management. Some of his work, however, (1959, 1964) is concerned particularly with goals, and the complex goal structures found in organisations. Thus he says (1964),

First, we discover that it is doubtful whether decisions are generally directed towards achieving a goal. It is easier and clearer to view decisions as being concerned with discovering courses of action that satisfy whole sets of constraints. It is this set, and not any one of its members, that is most accurately viewed as the goal of the action.

Whilst there appears to be an element of semantic confusion in this view (i.e. how in such a situation is a line to be drawn between goal and constraint?) it does reflect an important aspect of organisational behaviour, which it is intended to explore further later.

Another writer of the cybernetics of organisations is George (1970, 1974). He is one of the few people who, it can be maintained, has commented in depth on organisations from the standpoint of a profound knowledge of cybernetics. His main interest is, however, once again the solution of particular managerial problems through the application of cybernetic insight. Although he covers a wide field, from automation on the factory floor to major investment decisions such as diversification and acquisition, he pays little attention to the structure of organisations. The nearest approach to this general topic is when he discusses Executive Information Systems (1974, pp. 100-113), and then it appears he takes the roles and structures of management largely for granted. Thus, his introduction to the topic of information systems is as follows:

This chapter describes executive information systems, which are, generally speaking, a computerised version of data which is basic to decision making and planning.

It is quite vital to the success of such an information system that it be usable by senior management and easy for anyone to handle.

3.2 Feedback and Anticipatory Control

Perhaps the most relevant contribution to the particular aspects of cybernetics in relation to organisations of interest here is the work of Jankowicz (1973). He discusses management in terms of control and goal achievement. He identifies two types of control activity. The first of these is what may be termed "classical feedback", measuring deviation of output against goal and taking corrective action. The second is where control action is initiated on the basis of information of incoming disturbances reaching the manager via an input mechanism. This distinction is perhaps made clearer in diagrammatic form, as below (reproduced from Jankowicz).

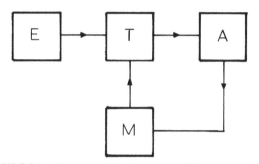

FIGURE 3.1. The classical feedback loop. (After Jankowicz.)

Figure 3.1 (Jankowicz's Figure 3) illustrates the "classical feedback" form. E is the environment, T is a transformation table, A is the manager's area of responsibility, and M is the manager.

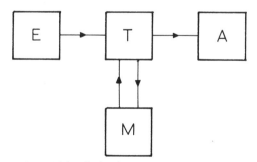

FIGURE 3.2. Jankowicz's modified feedback loop.

Figure   3.2 (also Jankowicz's Figure 2) shows the alternative form of feedback proposed by Jankowicz, where the manager is fed information direct from the input.

He then goes on to analyse the time delays inherent in such a system and points out that inevitably decisions will be delayed relative to the disturbances that they are designed to counteract, and some disturbance will be transmitted to *A*, and perfect control is not possible.

... the manager can only achieve control to the extent that environmental disturbances are not critical ... at every instant at which they occur; the same comment applies more generally to all feedback control systems.

Jankowicz apparently feels that this limitation on control is of serious consequence for an organisation, and proposes a type of control, "strategic control", to overcome the problem. The basic intention of strategic control is to reduce the time-lag in information reaching the manager (*M*), and is achieved as shown in Figure 3.3:

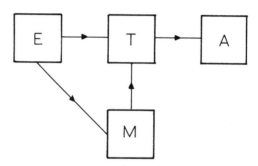

FIGURE 3.3   Jankowicz's proposal for 'strategic' control.

What this channel consists of is not specified. However, unless it is assumed that the manager has "direct awareness" of the environment (whatever that may mean) some form of encoding/decoding mechanism must be assumed in the channel $E \rightarrow M$, and it is difficult to see how this channel can then function faster than $E \rightarrow T \rightarrow M$ (for any principle that can be applied to $E \rightarrow M$ can also be applied to $E \rightarrow T \rightarrow M$).

However, this point is not essential to Jankowicz's main line of argument; if one considers the difference between "classical feedback" in figure 3.1 and the alternative form in figure 3.2, then this latter can be considered to contain the essential elements of "strategic control".

As conceived by Jankowicz, the nature of strategic control appears to be essentially predictive. Disturbances in the environment *E* are to be classified into two disjoint subsets, those which will be critical to the organisation and those which will not. The former subset, once identified, cause *M* to change its mode of operation. As Jankowicz himself puts it, the characteristics of strategic control are:

a) It acts as a parameter to individual control operations in T. Thus if we were to see M together with T as a finite automation, the parameter change involved in the $(E_1 — E_2)$ stage results in $M + T$ taking on new responses, coping with new disturbances, indeed becoming a different finite automation by changes in its transformations. The $M + T + (E_2 — E_2)$ stages thus constitute a finite function machine, rather than the "push-pull" finite automation $M + T$ whose functions (transformations) do not change over time.

b) As a parameter, it is at a higher level of discourse (acts within a higher universe of phenomena) than individual control operations.

c) It must impose some delay on the environmental disturbance.

Whilst the present author is in agreement with the fundamental concept of strategic (or predictive) control as a function of management, the formulation given above requires some comment.

In the first place, there is no detailed mechanism described which will enable the environmental disturbances to be partitioned into "critical" and "non-critical" subsets. This is assigned to the (somewhat mysterious) powers of senior management. In practice, such a distinction is by no means easy to discern. (For example, the appearance of Japanese-manufactured mopeds was not immediately obvious as a threat to the U.K. motorcycle industry, though it has turned out to be merely the thin end of a very long wedge.) Equally important, there is no mechanism suggested for selective attention to specific features of the environment ('perception' to use a psychological analogy), yet this is surely essential.

A further area where comment is required is the mechanism by which a parameter change is induced. This is apparently envisaged as a new version of T, where T is an Ashbean-type input/response/outcome table. Jankowicz does not suggest how a new table may be constructed, yet it must be assumed that a new table is required, or a parameter-change would not be needed. Nor can it be assumed that there is a store of T-tables available, ready for use. If this were the case, the situation would have occurred previously, and thus would be known not to be critical. Furthermore, if a store of T-tables were available, it would only put the question one stage back as to where they originated.

This leads on to the basic philosophical position behind Jankowicz's approach. He appears to see organisations as finite-function machines, i.e. as deterministic systems. This in turn enables him to construct tables (T) of required responses to produce a required output. It is doubtful whether, in practice, such a philosophy is applicable to real managers in real organisations. Outcomes of courses of action are difficult to predict with any confidence.

As a final point, Jankowicz asserts that strategic control "must impose some delay on the environmental disturbance". It is difficult to envisage how this may be achieved.

However, these difficulties with the proposed model should not obscure the fundamental point that is being made (indeed Jankowicz himself seems aware of some of the difficulties, though he does not include them in his formal model). This point is, to put it at its simplest, that organisations need to look to the future and act in anticipation of events, rather than just react to them.

Jankowicz also goes on to discuss the topic of self-organisation in relation to organisation, and models this in terms of information theory. Here he seems on less certain ground. He states that, "... for any system to increase its level of organisation over time, the rate of change of redundancy of its states should increase over time." Taking redundancy as

$$R = 1 - (H/H \max)$$

where H max represents the entropy of the total possible states of the organism, H the entropy of its states at any one time, we can derive the rate-of change inequality

$$(1/H_{max})\frac{dH_{max}}{dt} > \frac{1}{H} \cdot \frac{dH}{dt}$$

However, this will not ensure that the rate of change of redundancy will increase over time. What is required is that

$$\frac{d^2R}{dt^2} > 0,$$

which leads to a much more complex expression which it is not intended to examine here.

(It is also worth noting that the above treatment assumes that H max is variable with time. This would appear to be an arguable assumption — it could equally be assumed that H max is fixed for a given system. This leads immediately to the much simpler inequality

$$\frac{d^2H}{dt^2} < 0.)$$

Jankowicz then goes on to map H max onto the total variety in the T-table (the product set of environmental disturbances and reactions from M) and H onto the subset of T that satisfies the organisational goal-set, G. No justification for this mapping is given, and it is not intuitively obvious that it is correct. For example, it is not obvious why the total possible states of a system should be a function of the disturbances in the input to the system, yet this is what the mapping implies. Equally, the mapping ignores Ashby's concept of equi-finality, that a given result may arise from more than one state of a system. Additionally, no consideration is given to the possibility that G may itself vary over time.

In summary, although some of his conclusions are open to doubt, Jankowicz has pioneered a cybernetic approach to the nature of organisation structure. It is a topic which is well worth further exploration.

3.3 Neural Nets and Entropy

No discussion of the cybernetics of organisations would be complete without reference to the work of Beer (1959, 1962, 1966, 1967). Perhaps more than anyone else, he has developed the application of cybernetic ideas within organisations.

Beer's approach is derived basically from the discipline of Operational Research, and he sees cybernetics as one of a collection of scientific tools available for solving problems, rather than as the discipline best suited to the examination of the whole complex nature of organisations. This can be seen, for example, in *Decision and Control* (1966) where only one part of the book (Part III, Chapters 11-15) is devoted to cybernetics. Furthermore, Beer too is largely concerned with solving specific operational problems facing an organisation (How to control this machine shop? Where should a new factory be located?) rather than examining the more general problem of how organisations function and how they should be designed. Where he may be considered to be different from other writers is in his derivation of particular solutions from broad scientific principles. Thus *Cybernetics and Management* is at least as much concerned with expounding scientific philosophy and its relevance to management as it is with details of applications and results.

Beer's most detailed and explicit examination of the cybernetic aspects of management is to be found in *Towards the Cybernetic Factory* (1962) which consequently merits close attention. In passing, it is perhaps worth commenting that the use of the word 'factory' indicates a rather limited view of organisations, even of business organisations, taking no cognisance of equally important activities such as finance, selling, marketing, and so on. Beer's expressed view of management (i.e. ".. stock control, stores control, financial control, cost control and other functions of management . . ." (pp. 28-29) gives a rather limited range of activities, centred round mechanistic control procedures and does not cover the totality of the job outlined in

Chapter 2. It is perhaps also of relevance that Beer admits that the theory he presents was developed to account for a successful technique, rather than being the precursor of that technique. †

Beer's cybernetic account of a factory is in set-theoretic terms, and uses the analogy of a brain, "The cybernetic study ... went on to construct a model of the company organism and its environment and to detect the brain-like aspects of its control. " Much of the paper is concerned with developing a set-theoretic model of brain functioning, and it is a matter of some concern that the question of how this model maps onto the real-life firm is not examined. It is assumed, but not demonstrated, that such a mapping can be performed.

At a broad level, the brain model consists of a sensory mechanism (the T-machine) a decision-taker (the U-machine) an output mechanism (the V-machine) and a reward-mechanism (the R-machine, or 'algedonic loop'), which seems to be similar in many ways to a positive feedback loop. At this level of description, the model is unexceptionable. However, there are a number of unresolved problems when the more detailed model is examined.

Beer's initial model is of the T-machine, which analyses the input set G. (For typographical reasons the notation here does not always follow the original exactly; it has been rendered into Roman equivalents.) It would appear that T, which is a form of neural net, initially analyses the input elements Si ($Si \in G$) into "sensory configurations" via some kind of perception-like association process. Thus, section 1.2.4., "... the formal cortical networks generated by G, for which the ith elemental sensory input is either activated or not." This kind of model has been used for the brain elsewhere (See, for example, Stewart (1967); George (1961); McCulloch (1965)) and is again unexceptionable. The problem is that in parallel with this cortical network, Beer uses the concept of quantification of the inputs, "each input Si is assumed to be assigned a value Xi" but omits to discuss how the values Xi may be generated, stored, or processed. Furthermore, no evidence is presented that real brains work on analogue values of this kind. Yet, later in his discussion of the T-machine the use of such values (via a measure-set, Xn) is critical to the model. Complex transformations of the measure-set Xn are called for — eg. p.43:

1.5.1. The assumption is now made that the brain artefact will find some degree of statistical homogeneity convenient in its treatment of these numbers. To achieve this a succession of statistical transformations will be necessary. 1.5.11. There are various transforms (for example $y = \sin^{-1} \sqrt{(x/x)}$ that will tend to return a skewed distribution based on ratios to normal ....

---

† See *Decision and Control*, Chapter 13, p. 338. "As a matter of historical fact, the stimulus for the creation of the prototype system of this kind was found in production control. The methods described were devised in 1949 and 1950 for the solution of a practical problem; the full and more generalised account of the underlying theory .... was not achieved until later."

Presumably, it must also be assumed that knowledge of such transforms and the ability to use them, is inborn into the brain (it is difficult to see how they could be learnt, if the use of them is necessary to brain functioning), which argues for an extremely high genetic inheritance of structure, and again no evidence is presented for this. Nor is there any discussion of the neural networks required for such transformations.

There is a similar lack of discussion of another important aspect of the *T*-machine, or sensory cortex. In 3.25.1, p.59, it is stated ". . . the sensory cortex, (with its learnt patterns and ability to forecast) . . ." Nowhere in the formal description of the *T*-machine is there mentioned any ability to forecast, or how this may be achieved. Yet, this feature is crucial in the operation of the brain.

There are other difficulties with the model, associated with the amount of computation required. One example of this may suffice. Thus in 3.26.22, p.61, it is stated that "Therefore, the maximum structural variety . . . which converges on the *U*-machine is $2(2^{2n})$, and Beer seems to consider that n of the order of 30 is possible in practice, e.g. (p.66)." ". . . further experimental exemplifications have already brought the number of sensations considered in this work up to 36. . ."

Beer himself appears to be aware of this difficulty. For example, he says in 3.26.23, "The expression for the channel capacity required for output is elusive. . ." Or again, in 3.26.24, ". . . attempted calculations suggest, for example, that the transfinite cardinal must in practice be reduced to a cardinal of 4 or 5 . . ." But in 3.26.21, the value of this cardinal is given as $2^{2^{2^{2^{2}}}}$, where G is the set of sensory inputs. Putting these two statements together yields

$$2^{2|G|} = 4$$

and hence $|G| \simeq 1$ — a very limited set. This apparent conflict is not resolved in the paper.

A further point of interest is Beer's description of his *U*-machine as "an Ashbean homeostat" which is densely interconnected both internally and externally. He does not examine the problem of how long this machine would take to reach stability — or indeed, whether such stability is desirable.

˙However, in a sense, problems of the detailed functioning of the brain-model are not directly relevant to problems of management, particularly since, as noted above, the mapping from the model to the factory is not well specified. It is thus, in a sense, quite separate from problems of managerial cybernetics.

Beer goes on to discuss an exemplification of his theories in a practical situation (although, as has been pointed out, in fact the exemplification preceded the theory). On examination this exemplification appears to be chiefly, if not exclusively, concerned with the *T*-machine aspect of his brain-model, i.e. with statistical transformations of input data. This work would appear to be a highly successful and original approach to the design of a management information system. By using a series of transformations Beer succeeded in producing a highly relevant homomorphic mapping of input onto a set of predictive measures. Furthermore, he succeeded in making the mechanism of the mapping adaptive, to reflect changes in operating conditions, an advance whose significance is perhaps not generally recognised. It does not seem to have been followed up elsewhere.

Going on from this point, Beer's other work (1959, 1966, 1967) shows a great deal of concern with problems of variety and regulation (in the Ashbean sense). He asserts that organisations exist in an environment of extremely high variety, and is interested in cybernetics as a means of assisting organisations to cope with high variety. In particular, he is concerned with the concept of a "black box" inserted into control procedures to provide sufficient variety in the control loop to cope with the input variety, and with the relation between (thermodynamic) entropy and measures of information.

This approach is arguable as to its correctness. In the first place, if it is true that organisations need to cope with extremely high variety, then it is equally true that they do so successfully — organisations are extremely viable entities. It would seem more appropriate scientifically to attempt to establish what mechanisms are employed to cope with variety than to import mechanisms into organisations to achieve this end.

Furthermore, his more detailed approach to variety and requisite channel capacity seems confused. Thus, in *Decision and Control* (p. 252) he illustrates his point with a hypothetical set of 7 binary elements, in which all possible interconnections are allowed, which yields a variety of 2^{42} distinguishable states. Mapping this set onto a machine-shop with seven machines, he says "The manager has to handle a system of great complexity, it was said; just how great is the variety that must be handled is now beginning to emerge as a measured quantity." Later (p. 282) he relates this variety to the manager's task via Ashby's Law of Requisite Variety, e.g. he says, " ... the capacity to proliferate variety within the control box must be as great or greater than the capacity of the situation box to proliferate variety".

This view seems erroneous (or at least incomplete) on two counts. In the first place, Beer has omitted the important variable of time. Thus, in his hypothetical example, the variety generated is 42 bits; for control purposes, it is important to know over what span of time this total variety may occur. If it takes one minute for the system to permute over all its possible states, then the rate of information transmission is $42/60 \simeq \cdot 75$ bits/ second — which is by no means an impossible channel capacity for a manager to achieve. (In practice, one would assume that it would take much longer than a minute for a machine shop to pass through all possible states.)

In the second place, Beer appears to misinterpret Ashby's law. The Law of Requisite Variety establishes an upper limit to the amount of regulation or control that may be achieved; it does *not* state that control channel capacity must equal or exceed a situational rate of variety for any control to be achieved. In fact, the maximum amount of control that can be achieved is expressed by the difference between the two; if control capacity is less than situational capacity, there will be residual variety left in the output. From the organisational point of view, such a situation may be perfectly acceptable; production output may vary by ± 10% per day, but the situation is not critical provided there is sufficient storage capacity in the system and there is no long-term trend in the daily average.

There are two further points that are relevant here, concerned with the actual amount of variety generated in the environment. The first is that there are causal laws operating in the environment; knowledge of (or discovery or invention of) such laws will serve to reduce considerably the variety input to an organisation. The second is that variety is a measure imposed by an observer on a system, rather than an intrinsic property of the real system. Thus, to take an example from Beer, if the input to a system is billets of steel, the input variety is a function of the measures applied to such billets. If the measures are weight in milligrams, length in micrometers, chemical composition to .001%, then the input variety is likely to be high. If the measures are simply the number of lumps of mild steel weighing about 5 tons and between 18 and 22 feet long, the input variety will be correspondingly low. Following on from this, it can be seen that, in fact, organisations will themselves take measures to restrict the input variety to an amount with which they can cope; if it truly is necessary for the input billets to be accurate in length, weight and composition, the organisation will seek suppliers who can meet these specifications.

Thus when Beer discusses the need to introduce sufficient variety into the control system via a "black box" (see, for example, *Decision and Control*, Chapter 13. pp. 229-334), it is possible to question the logical basis for such a requirement. This is particularly so when it is realised that, on close examination, the effect of his "black box" is to effect a reduction in transmitted variety, not an increase.

Beer also discusses at length the relation between thermodynamic entropy of a system and information-content of a system. His starting point is the similarity between the equations

$$S = k \log g \qquad \text{(entropy)}$$
and
$$I = -\sum_i p_i \log p_i \quad \text{(information)}$$

This is a dubious base, unless some closer connections can be found between pi and g. To illustrate this, consider the equations

$$r^2 = x^2 + y^2 \qquad \text{(a circle)}$$
$$a^2 = b^2 + c^2 \qquad \text{(Pythagoras)}$$
$$\sigma^2 = \sigma_1{}^2 + \sigma_2{}^2 \qquad \text{(variance)}$$

Does this imply that a circle is the same thing as a right triangle, and that both of these are an experimental variance? Such a conclusion is not logical. Or further, consider the equation for intensity in dB

$$dB = 20 \log(P_1/P_\phi)$$
$$= -K \log P_1$$

Does this infer by analogy that the information content of a message is equivalent to its intensity? Again, such a conclusion appears peculiar.

On closer examination, the alleged equivalence between information and entropy appears to rest on a misinterpretation of the meaning of the variable g. Beer states its meaning (p.356) as "If the innumerable ways, of which there are, (say) g, are all equally likely to occur, then the entropy moves as the logarithm of g." In other words, g is the number of possible states of the system. On the other hand, Boltzmann's derivation of the entropy equation (as given in Allen and Maxwell (1952), pp. 815-816) assigns to g (given as W in the text) the probability of the most likely state of the system. These two meanings of g are substantially different, and the physicists interpretation must be accorded precedence. It is perhaps also worth noting that Boltzmann's derivation has been the subject of criticism, and that alternative expressions for entropy are available not involving the notion of 'number of states of the system' but based on physical dimensions such as energy and temperature. Furthermore, it may be of relevance that entropy as a concept is usually applied to closed systems, information to open systems.

Overall, it would seem safest to say that, although there may be a relation between entropy and information, such a relation has not as yet been satisfactorily demonstrated. Until it has been so demonstrated, there must be considerable doubt as to the reliability of any conclusions drawn from such a supposed relationship.

A further point raised by Beer is the application of Ashby's concept of homeostasis as a description of the interactions both between internal departments of an organisation and between an organisation and its environment. (See, for example, 1966, p.257, or 1967 pp. 156-162.) The organisation is modelled as attempting to come to equilibrium via a progression through unstable states until a stable set of interactions is reached, in a similar fashion to Ashby's Homeostat (see Ashby, 1960, Chapter 8). Although this parallel is in some ways attractive — it certainly reflects the constantly changing patterns of activity within an organisation — it is open to doubt whether or not it is an accurate account of organisational philosophy. It could equally well be argued that much of the functioning of organisations is designed specifically, conciously or unconciously, to avoid any permanent homeostatic equilibrium. Companies pursue a constant policy of innovation and change, in what can be interpreted as an attempt to veto any possible state of equilibrium. Indeed, it is probable that a company that achieved a policy of homeostatic equilibrium would be regarded as stagnating.

It is perhaps true that, in the short term, the combined effect of a large number of organisations interacting with the market produces a kind of apparent equilibrium in which variables such as market share and pro-fitability remain reasonably constant. History suggests, however, that these are comparatively short-term stabilities, as illustrated by the rise of new technology bringing obsolescence to many industries.

It is in fact arguable whether such a state of homeostatic equilibrium as is proposed by Beer is in fact desirable. It would seem that the most obvious exemplification of the results of such an approach can be seen in the early civilisation of Egypt. Certainly, equilibrium was established in that society — it lasted for millennia — but the result was complete stagnation, and eventually a slow decline. Progress and equilibrium can thus be argued to be opposed to each other, unless the equilibrium that is being discussed is of some highly abstracted variable.

However, the argument is now straying well away from the topic of the cybernetics of organisation. In summary, it can be said that the cybernetic study of organisation has been the subject of comparatively little attention. Most of the work that has been done has been concerned with solving parti-cular operational problems, rather than examining the nature of organi-sations and their management. The work that has been done in this latter field appears open to a variety of questions.

4

The Cybernetics of Management

What follows now is an attempt at an indepth account of organisations, and particularly the management of organisations, from a cybernetic viewpoint. It falls into four main parts. Firstly, a detailed model is developed based on cybernetic concepts. Secondly, the model is compared with the reported nature of organisations to see how it accounts for known aspects of organisational behaviour, and to validate it as a model. Thirdly, a practical application of the model to real-life situations is reported. Fourthly, the theoretical properties of the model are developed, to provide further insights into the needs of management.

4.1 A Model of Organisation and Management

A convenient starting point for building a model is with a systems engineering approach to the operational activities of the organisation. At a very broad level of detail, these can be mapped onto a system diagram such as that given in Figure 4.1.

FIGURE 4.1   A basic system diagram of organisational activity.

Block diagrams such as these can be expanded to much greater levels of detail, showing specific functions and information flows. Examples of detailed analysis of this kind can be found in Beer (1967) Strank (1971) Forrester (1961, 1968), Goode and Machol (1957) and several others. Such models, when appropriately quantified, have been found to be extremely useful tools for the analysis and design of operational systems. Generally, however, these models are not extended to include management operations as part of the analysis; at best, they indicate points at which decisions are required by (presumably) management, without examining the way such decisions may be arrived at.

Such diagrams can be extended, however, to give some indication of management activity. The justification for this extension lies in the fact of perturbation. In real life, the operations of an organisation will be disturbed by a variety of influences. Some will arise from within the organisation (e.g. machines will wear out or fail, employees will make errors) and some will arise outside the organisation (e.g. supply and prices of inputs will vary, c.f. Beer's environmental variety). In order for the organisation's operations to continue to run, a degree of regulation will be required, which, it is apparent, can be divided initially into two categories, internal and external.

Internal regulation as a term is intended to cover those activities which an organisation undertakes to adjust its internal operations to cope with perturbation, and can itself be subdivided into two categories, according to whether the disturbance originates as an internal malfunction or as input variety. These two subdivisions can be equated to Jankowicz's concepts of feedback control and strategic control, or (somewhat less precisely) to the managerial concepts of administration and planning.

External control as a term is intended to include those activities which an organisation may opt to undertake to achieve some degree of regulation over its environment. This is an area which has received little attention in the literature, but is a common form of organisational activity, and which can be broadly divided into three subcategories, the input environment, the output environment and the social environment. Organisations frequently take steps to regulate their input by applying a degree of control to their suppliers — for example, contracts may give quite precise specifications, several suppliers may be used to ensure continuity of delivery, and so on. An extreme example is where an organisation will purchase an outside supplier outright, which can be interpreted as an attempt to regulate its input. A different example is the case where organisations attempt to influence educational and training institutions, to ensure a supply of suitably qualified employees.

Organisational attempts to regulate the output environment can be grouped under the general heading of sales promotion and advertising. The intention here is clear, to regulate the market in favour of using the organisation's product and is often regarded as a key business activity.

Organisations exist within societies, and are influenced by the society in which they find themselves. Society as a whole attempts to regulate the organisations within it (for example, it legislates on certain activities that companies must, or must not, undertake). Equally, companies attempt to influence society, by asserting that their goods and activities are socially acceptable, and by forming pressure groups, to influence power centres within society, particularly government. This latter process is documented, for example, in the work of Gamson (1968), Olson (1965) Eckstein (1960) and Nettle (1965). Although they examine the process from differing viewpoints, they all agree that organisations bring influence to bear on governments to further their own ends.

Obviously, the extent of such activity will depend, amongst other things, upon the size of an organisation and the threat or opportunity perceived at any given period. The main point though is that organisations are involved in this type of regulatory activity, and a full cybernetic account of organisations needs to allow for it.

To summarise at this point then, on the assumption that potentially disruptive perturbations will arise both within and without an organisation, it has been established that there will be a need for control activities to ensure survival. Furthermore, there are two distinct categories of such activity, internal and external regulation, which can be further subdivided. Each of these merits detailed examination.

4.2 Internal Regulation

4.2.1 Feedback or administrative control

Feedback control can be illustrated on a block diagram as shown in Figure 4.2, which is a modification of Figure 4.1.

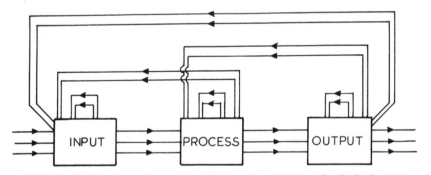

FIGURE 4.2.   Showing a schematic arrangement of control loops for the basic system.

This shows schematically the feedback loops required to control the effects of internal malfunction and random disturbance in the input. Several loops are shown, to illustrate that many variables will need to be controlled, not just one. Two loops are shown associated directly with each major functional area, to illustrate that each of these will have a number of variables to be regulated locally. Two further loops are shown connecting adjacent major functions (e.g. between 'input' and 'process') to indicate the possible need for co-ordinated action by two functions. Finally, two further loops are shown, covering the whole organisation, to indicate the possible need for co-ordinated action by the whole enterprise.

It will be appreciated that Figure 4.2 is schematic in the extreme. In practice, the functional organisation of an enterprise is more complex than the three-box approximation given, and many more than these two variables require to be controlled. However, such elaborations involve no difficulty of principle. A point that does require some comment is the justification for the control linkages between major functions. It could be argued that local control of each function should be perfectly sufficient to enable proper performance to occur; each function would maintain its output within limits to allow other functions to perform properly.

This argument, however, depends upon the assumption that the overall organisation, and each constituent function, has been properly designed and specified to fit into the overall system, in full knowledge of all problems likely to occur. This cannot be assumed to be the case (many instances could be cited where it is not — see Forrester (1961) for example) and thus there emerges a requirement for functions to interact via feedback.

It is not asserted that this requirement is necessarily fulfilled in actual organisations, in fact, as others such as Pask (1961) and Beer (1967) have noted, the normal hierarchical form of organisation structure puts great barriers in the way of such horizontal communication. It would seem possible to conclude on this basis that traditional management structure is based (consciously or unconsciously) on the premise that its systems are well designed. Clearly, if this were the case, and no horizontal communication was required, then the traditional management pyramid emerges as the proper organisational structure (at least in terms of internal feedback regulation).

To some extent, the argument is being anticipated here. What has been demonstrated is a need for a multiplicity of feedback loops to control an organisation. These can be identified as the task of management. In particular, a group of such control tasks can be brought together and assigned to one person, a manager. (Note that this does not imply that a manager is necessarily solely concerned with regulation of the enterprise; indeed the title of 'manager' can be bestowed more as a mark of organisational status than any necessary connection with regulation and control. In what follows, the terms 'manager' and 'management' will be intended to refer to control activities as set out above.)

It is also worth pointing out that however desirable it may be from a theoretical viewpoint, in practice there need be no logical connection between the individual control loops that are grouped together to form a task. Nor is it unknown for what is essentially the same control loop to be allocated to more than one person; perhaps the most outstanding example of this practice is the use of inspectors to check on operatives work, but examples of the same thing can be found at higher levels in the organisational hierarchy.

As a final note of caution, it cannot be guaranteed in practice that all the control loops that are theoretically required will actually exist in any given organisation. For example, many companies have found themselves in difficulties because of failure to install adequate control of cash flow. The reverse situation is also possible, in that organisations may install control loops that are either irrelevant, redundant, or even harmful.

It is also relevant to point out here that analysing an operation into constituent sub-functions is as much an art as it is a science. Different analysts, especially if they come from different backgrounds, are liable to arrive at quite different pictures. This applies just as much to a manager setting up an operation as it does to a scientist studying it. Yet this analysis will have a major effect on what control loops are installed. All in all, it cannot be assumed with safety that the controls found in practice are a unique — or even adequate — set.

With these points in mind, it is appropriate now to start to build up a more detailed model of feedback control, starting, for convenience at the lowest organisational level, that of the operative. Operatives jobs can be described by a feedback model, as discussed for example by Welford (1968). The basic nature of the model is as shown in Figure 4.3.

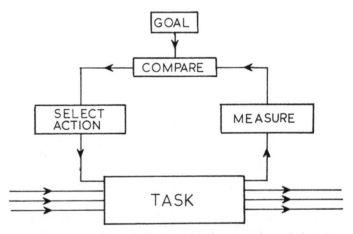

FIGURE 4.3.   A simple feedback model of an operative and his task.

The operative measures the progress of his task in some way (not necessarily via instrument readings) compares this with the goal of the operation, and selects an action designed to either correct any observed deviations or continue along the chosen path if there are no deviations. (This type of model is very common in a variety of contexts.) It is, however, deficient in at least one important aspect, in that it suggests that the operative has a single unitary goal. Even at operative level this is not the case, and a more representative diagram of the situation would be as shown in Figure 4.4, where four goals are shown.

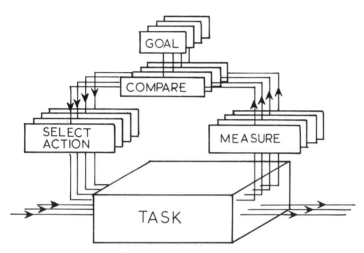

FIGURE 4.4. An expanded model of an operative and his task.

This aspect of organisation has been discussed by Jankowicz (1973) where he introduces the concept of a goal-set, G where

$$G = (G_1, G_2, G_3, \ldots G_j \ldots G_m)$$

and furthermore where each G_j may set up a series of subgoals,

$$G_j = (g_1, g_2, g_3, \ldots g_n)$$

which is in agreement with the formulation here.

It is perhaps worth re-emphasising that there is not necessarily any logical connection between any or all of the G_j. The common link may be only that they have all been allocated as the responsibility of a single person. Although it is rare to find a completely disjoint set for G, it is not uncommon to find that G can be sensibly partitioned into two or three distinct sub-sets —for example the job of telephonist/receptionist would break down in such a way. One consequence of this is that it can often be difficult to find a word or brief phrase that summarises the job adequately. Thus statements about the total goal vector G tend to be imprecise and nebulous, to the point where they become of extremely limited use as predictors of behaviour. Alternatively, as Pask (1969) has pointed out, the system can be regarded as possessing an underspecified goal.

It should also be pointed out that, due to imperfections in the design of the system, individual elements within G may be incompatible one with another, e.g.

$$G_5 \rightarrow \sim G_{20}$$

or perhaps in more complex forms such as

$$G_1 \cup G_7 \rightarrow \sim G_{19} \cup G_8$$

Incompatibles such as these are generally resolved by the fact that organisational goals are frequently in the form of ranges, (e.g. wages bill between £x and £y per week) or cut off points (e.g. return on capital not less than 12%) which gives sufficient room for maneouvre to approximate achievement of many goals. A further mechanism for the resolution of such conflicts is to assert that some elements of G are more important than others (this for example is a key assumption in the approach of Humble (1968)) and to concentrate on those. This approach may be considered as equivalent to attaching a weighting factor W_j, to each G_j, and the task then becomes one of maximising W, where

$$W = W_1G_1 + W_2G_2 + \ldots W_jG_j \ldots + W_mG_m$$

This is not necessarily a straightforward procedure if m is large and the elements of G interact, as suggested above. In practice, m may well be large; for example, a study of sales managers showed that each was responsible for 35 outlets, and had to control 15-20 quantifiable goals within each outlet, plus a number of qualitative goals. In this instance, m was therefore of the order of 1,000.

Another point of great relevance here is that the set G will contain the employee's own personal objectives as well as organisational goals. This is unavoidable, since it is impossible to employ a fraction of a person (which is perhaps the organisational equivalent of Planck's quantum theory?). Furthermore, Barnard (*op. cit.*) and Argyris (*op. cit.*) amongst others have agreed that some degree of conflict between individual and organisational goals is inevitable. Since they are not identical this conclusion is, of course, logical.

There seem to be two possible theoretical approaches to the resolution of such conflict. If G is partitioned into two sub-sets, organisational goals $G_j{}^O$ and personal goals, $G_R{}^P$, the first approach is to seek to maximise the intersection of these two sets, i.e. maximise P where

$$P = G_j{}^O \cap G_R{}^P$$

The second approach is to attach differential weightings to the various components of the G-vector, with the intention of using these weightings to persuade the worker that it is in his interest that organisational objectives should over-ride personal ones. Common examples of devices used to attach weightings to elements in G are wages, promotion, and continuity of employment. This can be formalised as attempting to maximise Q, where

$$Q = |W_j \cdot G_j{}^O| - |Wr \cdot G_R{}^P|$$

and W_j, W_r are the weighting factors.

These two strategies lie at the opposite ends of a continuum rather than being exclusive alternatives. In practice it is rare to find either in a pure form. A difficulty worth mention with the Q-approach is that the operatives' weighting-factors (the W_r) are outside the influence of the organisation. This may result in the factors on the other side (the W_j) having to be increased over the course of time.

These two approaches seem to correspond to distinct managerial styles. The P-approach corresponds with the human-relations school of management, as discussed by people such as Argyris (op. cit.) Mazlow (op. cit.) and others, allowing people to participate in the running of the organisation and aiming to create job satisfaction. The Q-approach would seem to correspond with the authoritarian school of management, whose outright proponents are not well represented in the literature but are characterised by McGregor (q.v.) as upholding Theory X.

It is of interest to explore the probable results of the two approaches to conflict-resolution. Using the P-approach can reasonably be expected to lead to a greater degree of worker involvement with the job, greater loyalty and better job performance. It will also involve recognition, implicit or explicit, that employees will have an influence on the goals of the enterprise, which may well be psychologically unacceptable to senior management. Furthermore, the gaining of psychological acceptance of organisational goals as overlapping with personal goals involves what may be seen as an act of leadership. The need for leadership will be higher in organisations adopting a P-approach, with a consequent need for greater personal belief in commitment to organisational goals on the part of senior management. This element of personal belief will tend to make it harder to make radical changes to organisational policy if they are needed.

One type of organisation of especial interest from this point of view is a Trade Union. In theory at least — and to a considerable degree in practice — the goal setting process is a reversal of the usual process. The goals of Trade Union leadership are determined by the common goals of individual members — and in this instance the goals will be personal rather than organisational. If the individuals personal goal-set is I^n, where

$$I^n = (I_1{}^n, I_2{}^n, I_3{}^n, \ldots I_1{}^n \ldots I_z{}^n)$$

then the goals of a Trade Union, I, can be symbolised as the intersection of the individual I^n, i.e.

$$I = I^1 \cap I^2 \cap I^3 \ldots \cap I^n \cap \ldots \cap I^r$$

In practice, I will probably reduce to a very small set, representing the comparatively few interests held in common by members. This difference in goal-setting structure has a profound effect upon the nature of Trade Union activity, and particularly upon its leadership. This is frequently forgotten, because the organisational structure of such bodies is outwardly similar to company structure.

The Q-approach is substantially different from the P-approach. It is essentially a bargain struck between employer and employee, where the latter agrees to subordinate himself to the former in return for some form of consideration, usually financial. (It is worth noting that the value of such a bargain to the employee depends upon his having time available away from the organisation in order to enjoy such benefits.) It is this type of approach that is likely to appeal to the entrepreneur stereotype, who conducts his affairs in this way. It can be expected to result in a great deal of concentration upon the heavily-weighted elements (usually wages) with the employee trying to maximise its value, the employer trying to minimise it. Hence, it can also be expected to encourage co-operative (or union) activities amongst employees. Furthermore, it does nothing to encourage psychological acceptance of organisational goals by employees.

Although there are no firm data to support this analysis it corresponds to subjective impressions of different types of organisation. Furthermore it facilitates cybernetic discussion of a range of problems reported in the literature (see Argyris (1960), Mayo (1933), Jaques (1961), Bernard (1948)) that have not received cybernetic attention.

Two aspects of the multiple-goal situation are of interest from a cybernetic viewpoint. The first is that bringing about a stable situation where all goals are met is a difficult problem in itself. Ashby (1960, Chapter 20) has examined an analogous situation, and concluded that the probability of stability of a multi-variable system decreases as the number of variables increases. He hypothesises that the probability falls off as $(\frac{1}{2})^n$ where n is the number of variables. Thus introducing a new variable into a situation, or changing one goal among many, can be expected to change the stability of the overall system dramatically. Furthermore even if a new stable region is discovered, its characteristics are likely to be markedly different from the previous situation. An intuitive appreciation of this may lie at the root of the phenomenon of resistance to change.

The second aspect, which is related to the first, calls into question the validity of attaching weighting factors to goals. Adopting such an approach may allow some variables to depart widely from desired values, which may in turn affect the overall stability of the total system. Such affects have been discussed at length by Ashby (1956), where he shows how a system may change abruptly from one field of behaviour to another when its state-vector exceeds certain limits.

Overall the picture emerging of organisations up to this point is one of virtually total instability, of constant teetering on the edge of violent upheaval. To counteract this, it must be borne in mind that the day-to-day operation of most organisations contain inbuilt inertia — particularly so in the case of large scale manufacturing operations.

So far, the examination of organisations has not proceeded very far. It is still at the operator level. However, the problems discussed at this level apply equally at other levels, and it is worthwhile to point out that the application of cybernetic principles can be made at all levels of organisation.

As far as administrative or feedback management is concerned, an organisation can be regarded as a series of hierarchically arranged supervisory feedback loops. Bearing in mind that G is a vector, the basic arrangement is as shown in Figure 4.5. where Gs is the supervisors goal-set, Gop the operators goal-set. The basic functioning of this supervisory loop is that the supervisor assess current performance against his own goals, and selects any corrective action required.

The most fundamental point about Figure 4.5, is that corrective action is not applied directly to the task, but to the goal-set Gop (this is encapsulated in the definition of management as "Getting things done through people"). Furthermore, in the present context, this corrective action is not basically a servo-mechanism type designed to track a changing goal, but is more akin to a simple control loop.

The question naturally arises as to whether there is any theoretical need for supervision of this type? Once the operator has accepted a set of goals, why should there be any need to check on his performance?

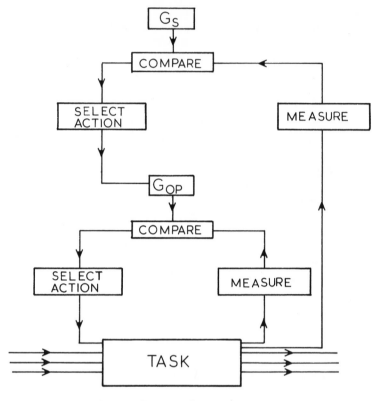

FIGURE 4.5.   A basic model of the supervisory control loop.

The initial answer to this comes in three parts. The first is that *Gop* is stored in human memory, it is therefore liable to decay and error; therefore, some reinforcement will be required. The second is that, as discussed above, *Gop* contains personal ambitions which will evolve with time. In turn, these evolving aims will, if unchecked, influence task performance. The third is that the organisational subset of *Gop* is derived as a subset of *Gs*. Thus if

$$Gs = (G1s, G2s, G3s \ldots\ldots Gjs \ldots\ldots Gms)$$

then

$$\overset{\circ}{Gop} = (Ges, Gfs, \ldots\ldots Gns)$$

and where each *Gfs* can be viewed as generating a series of sub-goals,

$$Gfs = (g1, g2, g3 \ldots\ldots gu)$$

(see Jankowicz (1973), Pask (1969)). Now if, as is likely, these goals and sub-goals have been set less than perfectly to bring about the desired results, then they will require to be reviewed and revised in the light of experience. This is a third function of the supervisory control loop, where overall purpose can thus be seen as compensating for the inevitable shortcomings of real people in real situations.

Having accepted the need for supervision of this type, it should not be imagined that the cybernetics of the process are as straightforward and simple as might be inferred from Figure 4.5 i.e. the straightforward issuing of an instruction which is promptly put into practice.

In the first place remembering that the corrective action is applied to the set *Gop*, the superior will find it useful to know what this set consists of — or at least, what are some of the major components. He is thus involved with the general problem of establishing the goals of a working system. Pask (1969) has examined this general situation, and indentifies two different strategies, either to observe behaviour and infer goals from the relation between input and output, or alternatively to enquire directly of the system what its goals are. Pask states this distinction as being between the system being regarded either as "taciturn" or "language-orientated", which distinction is basically a choice made by the observer rather than a characteristic of the system. (Though there are some systems for which it is difficult to discover the appropriate language.)

It is not proposed to pursue the point in great detail here. It should be noted, though, that either strategy can give rise to difficulties for the supervisor. Inferring goals from behaviour can lead to error, and equally asking the operator to state his goals can result in inaccurate or untruthful responses. Many supervisors in fact use both strategies together — and may spend much time trying to resolve the discrepancies between the answers from the two approaches.

A point of particular interest made by Pask is that, "A taciturn system can neither be given new goals nor can it state its goals." (Although, as he also indicates, it is possible to change parameters of a given goal within a taciturn system.) This is of relevance at a later stage in this model of organisation.

Another of the problems is that of language; instructions are issued and received via the medium of language. What is important here is not so much some of the deeper theoretical issues, as discussed for example by Chomsky (1957) or Morris (1946), but the more pragmatic aspects of the subject, as discussed by McGregor (1960), Mazlow (1965) and Drucker (1970). What seems to emerge from these writers is that it is necessary to set up and maintain a language that is meaningful to both parties — and furthermore that insufficient attention is paid to this problem by organisations. The results are commonly misunderstanding, misinterpretations and mistrust. It would seem from this that many managers adopt what could be termed an "information-theoretic" approach, rather than a "communication-theoretic" approach. They ignore the fact that it is meaning that is passed on, not simply information, and this can only be done in the framework of a language that has a common significance to both sides of the conversation.

The setting up and maintaining of such languages involved a good deal of continuing effort, and involves both sides of the conversation. In conventional management terms, the "upward" flow of the interaction should be of equal importance to the "downward" flow, as Drucker (1970) is at pains to emphasise. Furthermore, the setting up and maintaining of language is an essentially "off line" activity, which is facilitated by informal contact outside the working situation. It is of interest that many organisations actively discourage informal contact between different organisational levels. The basic result of such strategies is well discussed by Machiavelli.

On the face of it, the foregoing line of argument is refuted by the military situation, where the paradigm is orders crisply issued and instantly obeyed. It is worthwhile, therefore, to examine this situation in more detail.

On closer examination it becomes apparent that, in fact, military organisations do expend a great deal of effort to build up and maintain a language sufficient for their purposes. In this context, it is worth noting initially that the bulk of the armed forces' time is spent in "off line" activities — i.e. real (rather than simulated) combat is a relatively rare activity. Additionally, a code of discipline is rigorously inculcated and maintained. Military forces will go to great lengths to maintain dicipline, up to and including execution by firing squad. (This latter is a sanction not normally available to industrial management.)

The language used in military situations is highly codified, with exact terminology and usage; e.g. "Present arms!" is an order uniformly and universally interpreted, as a result of intensive training and discipline. It is also a relatively simplified language, which is perhaps a result of the G-vector generally being simpler in structure. There are fewer conflicting G-elements (for example, questions of cost seldom figure largely in combat decisions) and priorities amongst the G-elements are much clearer.

The importance of such matters in a combat situation, where troops may suddenly come under the command of an unfamiliar officer, can be readily appreciated. That is not at issue here. What is less clear is whether the lessons learned by the armed forces can be readily transferred to the industrial situation.

Thus far, communication (as distinct from information transfer) has been discussed and its importance to the functioning of organisation established. Information transfer is, of course, also an important function, and is represented in the diagram of Figure 4.5 by the processes "Gathering information", "Compare" and "Select action". The distinction can be illustrated by the fact that the first two of these can be (and often are) automated to some extent, frequently with the use of computers. It should not however be concluded from this that managerial information systems can consist entirely of various types of statistical report, such as profit and loss accounts. More information than this is needed for successful management.

There are two aspects to this question. In the first place, many important aspects of an operation cannot readily be quantified (at least, in the present stage of technology, though it is feasible that some progress may be made). Examples of such variables are such things as motivation and morale. Fundamentally, the information required will be determined by the G-vector; in principle, each element of G will require a feedback path to control it. Whether the necessary information is readily quantifiable or not will depend upon the nature of each particular G.

The second aspect of the question is more fundamental. In order to control a system, a model of that system must be provided in the feedback loop. (See Ashby and Conant (1970)). As far as simple feedback systems are concerned such modelling may be at a very primitive level (e.g. in a cistern, water level is modelled as the height of the float). Furthermore, as Ashby and Conant (*op. cit.*) point out, at this level of sophistication, "almost anything may serve as a model of almost anything else". Thus, if a manager takes decisions on the basis of "What would Uncle Fred do in a situation like this?", then Uncle Fred is serving as a model of the system under consideration. Nor, let it be stressed, is this necessarily a bad model *in the circumstances being considered here*. If "Uncle Fred" as a model yields good decisions, then there is no need to seek further.

However, the point to be made here is that the control model used does not appear, by some mysterious process, out of thin air. It is obtained by study of the system under consideration. Ashby (1956, 1960) has discussed the general problem in cybernetic terms and Garner (1968) provides specific examples of the modelling of human performance. Though neither writer specifies it in these terms, what is essentially required is a metalanguage to describe the system, to propose hypothesised variables and parameters, which are then used to experiment with the system and see if the hypothesised model is adequate. Thus management information also needs to cater for this need to set up models of the organisation, and this precedes any flow of information about values of particular variables. (The distinction is akin to the distinction made by Mackay (1950) between metron and logon content of information.) In practice, these extra sources of management information may be obtained by a variety of means, such as written reports or actual visits and physical inspection of the system.

It should not be imagined that the flow of such types of information is simply a once-off affair. Many managers are constantly updating their model of the organisation — indeed, the desk-bound manager concerned only with "the figures" is an archetypal whipping-horse in management training. He rapidly becomes divorced from reality, i.e. his model becomes inappropriate. The managerial model is vital to the interpretation of statistical information; without it, all the figures in the world are meaningless. In passing, it should be noted that the managerial model will contain a model of human behaviour, i.e. the system modelled contains a human element.

It should also be noted that these flows of modelling information are as liable as any other channel to noise and distortion. In practice, these factors may be deliberately introduced by the system under study; the consequent problems for management need no elaboration.

Thus it can be seen that the process "Gather Information" is not necessarily as straightforward in organisations as might appear from Figure 4.5. The extra considerations can be shown schematically as in Figure 4.6.

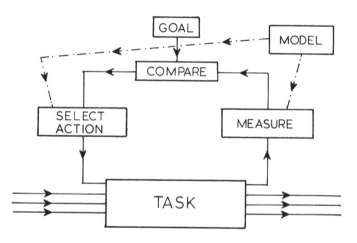

FIGURE 4.6.   Illustrating the influence of a managerial model on the selection of control information.

This indicates that the gathering of information may be viewed as a filtering process, selecting only those items which are relevant to the G-vector in question. The model provides the setting of the filter, and changes in the model can thus influence the data gathered. It is perhaps worth mentioning that a major type of model frequently encountered in organisations is the accounting or financial model, which is generally assumed to be constructed to reflect the working of the business. This is not necessarily the case, and it is not unknown for organisations to be reconstructed to fit a particular accounting model especially when accountants gain considerable authority in an organisation.

It is also worth mentioning that, although the actual gathering of data may be a reasonably continuous process, the issuing of reports and statistics is generally done at discrete intervals (a week, a month, a year). Thus a supervisory control loop is basically an intermittent rather than a continuous process.

After information has been gathered the next stage in the process is comparison with the goal. This again is not altogether a straightforward process. The difficulty arises when as is frequently the case, the G-vector is not well specified (seen in, for example, Humble (1968)).

Three particular types of difficulty can be identified. In the first place, it can happen that a control variable is specified, but no value is attached. Statements such as, "Manufacture item X at minimum cost" or "Deliver goods as quickly as possible" are examples of this. In instances such as these, the manager himself will supply a value believed to be appropriate — and the value chosen may well differ significantly from the value implicit in the mind of the manager's supervisor.

In the second place, although a value may be specified, no tolerance, or permissable range, is supplied. If for example, production costs are 2.4% above target, this information on its own is not sufficient to decide whether this is a minor inconvenience or a major disaster requiring a crash programme to rectify it. Here again, in the absence of other guidance, the individual manager will set his own tolerance limits. Variations on this theme are possible, such as a goal in terms of a limit function (e.g. labour turnover less than 10% p.a.) or a trend function (e.g. to reduce labour turnover).

However, it is the final stage, that labelled "Select Action" which is generally considered to be essentially a managerial function, usually under the title of decision-taking. (The term "Select Action" is preferred here, in an attempt to emphasise the importance of actually doing something as the result of a decision.) Other parts of the control loop can be (and often are) performed by others; it is the taking of decision that is the core of the managerial role.

Despite what has been written by others (e.g. Simon (1958, 1960), Kaufman (1968), Luce and Raiffa (1957)) it is the contention here that, *in the context of simple feedback management,* decision processes are, in principle, extremely straightforward, and do not need to involve the complex evaluations often discussed. In principle, all that is required is a simple black box model linking a deviation in the result of a task with the required adjustment to the input. There is no need to establish cause and effect, no need for complex evaluations of outcomes and payoffs. A simple black box approach will serve equally well, if not better.

(This does not of course imply that managers necessarily use such a model for control. They may use considerably more complex approaches. Observation does suggest that many managers do in fact develop a highly summarised view of the operations that they run. They concentrate on a comparatively few variables, and appear prepared to let the details look after themselves. These are the characteristics that would be expected from the use of a black-box model, as will be discussed later. It can also be seen how this phenomenom can lead to the inference that there exist "key objectives" (see Humble, *op. cit.*). How far this inference is justified depends on the extent and nature of the interactions between the variables in any given situation. It must also be remembered that these interactions are liable to change in the course of time as the operation evolves.)

Two complications of this basic contention deserves some discussion. The first is that, in the managerial situation, one option open when performance deviates from target is to change the goal. Since, as discussed above, goals are frequently not well specified, this often does not present much practical difficulty. It can be considered the organisational equivalent of the game-theoretic solution of "leaving the field".

The second is the complication introduced by the fact that G is a vector, and that the elements of G interact in the sense that an adjustment to return Gj to target may induce changes in the system that will disturb Gr. This is equivalent to saying that, when controlling Gi, the set $G - Gi$ acts as a constraint on the permissable actions. Thus it can be seen as suggested by Simon (1959) that a given variable can act both as a goal and as a constraint — but not at the same time; which depends upon circumstances at the time.

Up to this point, the cybernetic aspects of the supervisor-single operative situation have been considered. This does not correspond with the reality of organisational relationships (except in a few anomalous instances). In practice a manager generally has several direct subordinates — typically 5 or 6, though the range is from 2 to 40 or more. The model can easily be extended to show this feature of organisation, as in Figure 4.7. Here three subordinates are shown, though it can be seen that in principle the diagram could be extended to any number of subordinates. For the sake of clarity, the modelling information channels discussed above are not shown; for similar reasons, the supervisory control loop is shown as receiving all its information from the final output, though in reality information could be — and frequently is — derived from any intermediate point as well. The managers area of responsibility is defined by the points at which his information channels start and finish (c.f. Miller and Rice (1967), and the concept that the executive functions at the boundaries of the organisation).

Figure 4.7 shows a simple serial relationship between operation T_A, T_B, and T_C, typical perhaps of a large-scale flow process. In practice, much more

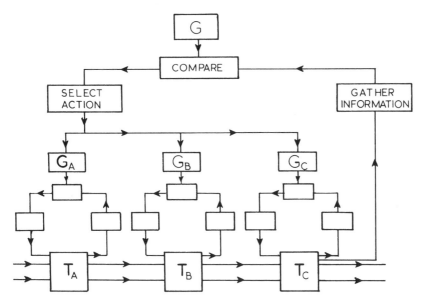

FIGURE 4.7.   An extension of the simple supervisory control loop.

complex relationships may hold between operations, and modelling techniques have been developed for such cases (see Forrester (1961, 1968) or Beer (1967) for examples). A case that requires special mention is where the supervisor is in charge of essentially parallel situations, such as a sales manager in charge of a team of salesmen, each with his own territory. This represents the opposite extreme to the series situation shown in Figure 4.7. Mixed series/parallel situations are, of course, possible.

This distinction between serial and parallel management situations would appear to be an important one, but one that has not been commented on in the literature, at least not in these terms. The major difference involved would appear to be in the type of model required in the supervisory control loop. In a series operation, the model must encompass a greater degree of complexity, a greater degree of interaction amongst the variables. In a parallel situation, the same basic model can serve for all functions — although it is desirable that it be given at least some degree of "fine tuning" to adjust it to the individual characteristics of each.

As well as the serial/parallel distinction, the introduction of several subordinates into the situation brings with it extra features of interest. These arise from the possibility of the operators or subordinates communicating among themselves.

Figure 4.7 can be modified along the lines indicated in Figure 4.8 to show some of the possibilities.

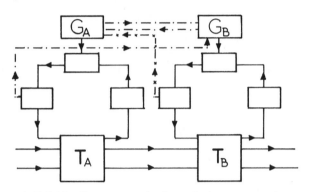

FIGURE 4.8.   Some communication paths between operatives.

Two main types of communication are indicated. One is a channel from the "Select Action" process of T_B (or T_A) to G_A (or G_B). This represents the situation where one operative decides that the best way to cope with his difficulties is to modify the operation of some other function (e.g. "Slow down a bit Fred, and give us a chance"). In organisational theory, such requests should go through the supervisory control loop; in practice they often do not. The other type of communication is a channel direct between the two G-vectors of the operatives. This represents the informal communication that will take place between colleagues, and will concern particularly, it can be presumed, those elements of the G-vectors that are personal rather than organisationally derived. Such communications may generate what are effectively new elements in the goal-set $G_A{}^P$ (particularly when it is remembered that the personal component of $G_A{}^P$ is subject to influences derived from outside the organisation).

This form of communication will obviously lead to the formation of working groups. This may result in mutually beneficial change — or equally lead to conflict and "clashes of personality". The generation of new elements in the G-vector can also give rise to what can be termed with some justification "organisational psychology", i.e. those aspects of behaviour arising specifically from the nature of organisations.

The arrangement shown in Figure 4.8 is equivalent to Pask's (1971) concept of minimal structure required for a meaningful conversation, if the inputs from T_A to T_B are regarded as low level language. This in turn implies that the system has the capability of self-organisation. (Though the same does not necessarily apply if T_A, T_B operate in parallel.) This may or may not work to the advantage of the organisation, depending upon whether the self-organisation is centred upon organisational goals or personal goals. This, in turn, will be a feature of whether the organisation uses the *P*-approach or the *Q*-approach described above.

Mapping the *P*-approach onto the human relations centred school of management and the *Q*-approach onto the authoritarian school, thus leads to the hypothesis that a human-relations centred organisation will tend to be flexible and adaptive in behaviour with comparatively good industrial relations but comparatively lacking in "business drive" (due to the fact that personal goals and organisational goals influence each other), whereas the authoritarian school will tend to be rigid and inflexible, have many disputes, particularly over wages, but will have a hard, aggressive approach to business, due to the dominance of organisational goals.

Data would be needed to confirm or deny this hypothesis, although it does appear to have some degree of face validity, at an intuitive level. Which form of organisation is superior is yet another question which would be of interest to answer (and indeed, remembering that organisations are fundamentally for fulfilling human needs, the criteria for "superiority" are not self-evident; simple measures of profitability are only part of the answer). The main point here, however, is that issues such as these can be seen to arise from cybernetic mechanisms and can be discussed in cybernetic terms.

It is also possible to attempt to quantify the likelihood of significant Trade Union or similar activity. This can be posited to be a function of, amongst other things, size of group, time spent on intra-group communication, and time spent in communication with a supervisor. Thus the probability of Trade Union activity *Ptu* can be expressed as:

$$Ptu = f(n, t1, t2)$$
Where n = number of people in work group
$\quad t1$ = time spent with work as a cohesive unit
$\quad t2$ = time spent in direct communication with superior.

It is possible to speculate on the possible form of the function. Thus, it can be expected that the dependence of *Ptu* upon *n* will not be linear, and may well be some form of power law, possibly quadratic. Since *Ptu* is a pure number, t, and t_2 must appear as a ratio. Therefore, an initial approximation to *Ptu* would take the form:

$$Ptu \propto n^2(t_1/t_2)$$

though experimental evidence would be required to verify this.

Even in such a crude form, the expression indicates that after factors being equal the probability of Trade Union activity is highest with a large workforce whose tasks interact and who therefore communicate often and where there is a low ratio of supervisors to operatives. This does not sound unrealistic as a reflection of the real world.

Another corollary of intra-group communication is that it enables subordinates to construct a far more comprehensive model of the supervisor, via shared experience, than the supervisor has of any individual subordinate. It is thus possible that, in appropriate circumstances, subordinates are better able to regulate some aspects of supervisory behaviour than the supervisor is able to regulate subordinate behaviour.

Thus far, the consequences of the possibility of self-organisation for the supervisory control loop have not been pursued. One feature of importance is that the control model used in this loop should, in principle, be revised to conform to the altered system it is trying to regulate. As discussed previously, this implies more than can be achieved through reporting systems. The structure of the model needs to be changed, which can only be achieved via the flow of modelling information. In practice, it is not uncommon for managers to complain that they do not know what is actually going on in the organisation under their control. It is important to realise that this is not necessarily equivalent to a statement that they see their area of control as a "black box"; it may imply that not only is it a "black box," but a "black box" whose input-output relationships are not static. It would appear that some managers find such a situation unmanageable and insist on standard procedures — i.e. actively inhibit self-organisation of their area of command. Others find it acceptable, and even encourage it. It may be that the root of this difference of attitude lies in the differences between the control model used by different managers (for nothing said up till now implies that there is any unique, or even optimal, model for simple feedback regulation).

Using Blake's (1969) dimensions of managerial attitudes (i.e. broadly "people-control" or "task-control") it can be hypothesised that a "people-control" manager uses models of his subordinates for control, a "task control" manager uses models of the operation for control. The former will be less affected by self-organisation, and thus such a manager will be more flexible and still maintain control. Furthermore, the operation under his command is likely to be better adapted to prevailing circumstances, and performance will be superior. Such a conclusion is supported by the work of Argyris (q.v.).

A related issue here is that in the normal course of events a manager may expect promotion and/or transfer in his career. In theory, this would imply that his models should be discarded and new ones built. In practice, it may be that the manager retains his models, and attempts to reshape his area of authority to conform with them. In practice, some middle course between the two may be adopted. Again, it can be expected that a manager with "people control" models could transfer more readily and painlessly than a more "task-control" manager. Thus it is not unknown for a highly competent manager within a technical specialism to be unsuccessful outside his own specialist field.

The diagram of Figure 4.7 shows a single level of management. It will be readily appreciated that the diagram could be extended vertically, to show a further control loop spanning two or more supervisors, and so on, which would then correspond to the familiar hierarchical model of organisation.

Only a few extra features of significance arise from such a vertical expansion, and no great discussion is required.

In the first place, it should be re-emphasised that a diagram such as figure 4.7 is not intended to imply that all the (theoretically) necessary control loops are in fact present in any given organisation or, vice versa, that control loops found to be present in actual organisations are theoretically necessary, or indeed desirable.

An equally important feature that has not been mentioned previously is that it is not necessarily the case in practice that all the control loops pertinent to a particular function are channelled through a single individual. (A case in point is the personnel department in many organisations; it is often the practice for personnel to legislate over variables such as hours of work, payment, and so on, taking the control of such goals out of the hands of the individual manager.) The general consequence of such practices, in theoretical terms, is that the control of a single function is mediated through two or more distinct models. These models may not necessarily be compatible one with another. The behaviour of this type of system does not appear to have been considered in the literature from a theoretical standpoint.

As a final point, it can be seen that expanding the diagram of Figure 4.7 will allow much greater opportunity for intercommunication in an organisation, and consequently great opportunity for self-organisation. The degree to which this self-organisation can occur can be influenced greatly by the managerial level to which particular control loops are routed; if many control loops are channelled through senior management levels, then the possibilities for self-organisation at the lower levels are correspondingly reduced. As the responsibility for certain control loops is passed to lower managerial levels (i.e. as delegation occurs), so the opportunities for self-organisation increase.

Such a process is similar to what would be described as "decentralisation" in traditional management terminology ("Centralisation" is obviously the reverse of this process.) What is of interest here is not simply that such a concept can be modelled in cybernetic terms, but that some underlying rationale for it can be discerned. Thus, if an organisation is in a reasonably static environment, and is not contemplating any fundamental change in its own operation, then it may well make sense to allow the individual parts of the organisation to attain local equilibria through self-organisation, i.e. to decentralise. Conversely, when co-ordinated action of the enterprise as a whole is required, to respond to either external threats or internal innovation, then it may be appropriate to centralise control.

It is not claimed that actual companies, do always centralise or decentralise for this reason alone; it may be undertaken for a variety of other reasons. What is of interest here is the possible logical justification in cybernetic terms of a well-known feature of organisational behaviour.

However, mention of centralisation and decentralisation brings up a key feature of the nature of organisations. This is that they have the capability (and frequently use it) to change themselves to meet new needs. This change can be at any level, and includes the ability to change management information systems and decision procedures — indeed, to completely re-structure any or all of the management of an organisation. There would seem to be no adequate analogy with this process within the natural world. It would be rather as if an organism could spontaneously generate a new type of input sense-organ and nerve structure for each environment in which it found itself. This is a feature of organisation that has not received much comment (apart from writers such as Burns and Stalker (*op. cit.*)). It is as though the approach to organisation has been based on the belief that there is one optimal form of organisation, and what is needed is research to identify it. However in cybernetic terms, the ability to change organisation can bestow great benefits in a changing environment. Indeed, it may be this ability that enables organisations to survive in environments that are arguably of much higher variety than environments that one is used to considering: there are few natural redundancies (laws of nature) in the organisational environment. In fact, the environment of organisations can be regarded as made up almost entirely of other organisations.

This line of argument is leading on to topics that are more readily considered under the heading of strategic management. Before passing on this topic, it would be as well to summarise briefly what has been discussed up to now.

A feedback model of management has been proposed, and developed in detail. This has been found sufficient to explain many of the reported features of organisation, and offers the means of quantifying many of the problems occurring in organisations. It has been concerned basically with the problems of line management.

4.2.2 Feedforward or strategic control

In addition to the feedback mechanisms described previously, it seems necessary to hypothesise a further set of basically anticipatory (or what may be termed feedforward) control mechanisms. The basic reasons for postulating such mechanisms are cybernetic necessity on the one hand and observed management practice on the other.

The basic form of such a mechanism is well known; it was proposed by Ashby (1956) as the basic model for regulation. As adapted for the purpose here, it can be shown as in Figure 4.9. The basic principle illustrated is that, instead of using information about output behaviour, strategic control uses information derived from the input to adjust organisation performance. Ashby (1956) showed that this was canonically equivalent to the more normal feedback characterisation of control activity.

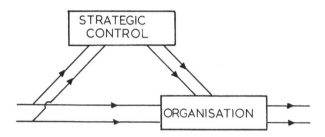

FIGURE 4.9.   The basic information flow in strategic control.

As has been discussed above, Jankowicz (1973) has used this model for the analysis of management, and indeed the term "strategic" has been taken from his work. He apparently saw the chief virtue of strategic control as reducing time-lags inherent in a feedback system. Thus he says,"The second type of control, *strategic control*, attempts to regulate disturbances by reducing the time in which information reaches the manager." Furthermore, he sees this type of anticipatory action as the role of senior management, as setting parameter-values for the operation of lower-level managers.

These arguments do not seem to cover all the important aspects of the nature of strategic management feed forward control. As far as speed of response is concerned, although the delay involved in feedback is widely recognised, in practical terms it is usually not of any great consequence, particularly if the delay is small compared to the rate of change in the environment. Furthermore, the feedback process itself, if suitably elaborated, can provide a sufficient framework for parameter changes within the system, as has been described.

Indeed, on closer examination, the idea that feedforward control necessarily facilitates regulation by improving speed of response is not so simple as it at first appears. It is worth pursuing this point in some detail, since it leads on to clarification of important areas of management activity.

The diagram of Figure 4.9 can be expanded as in Figure 4.10 to show the nature of the feedforward loop in greater depth.

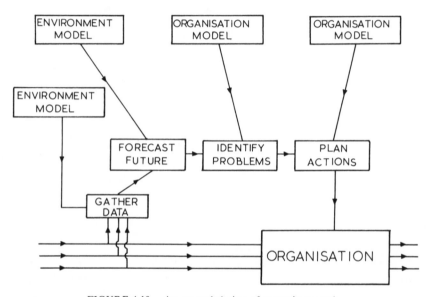

FIGURE 4.10   An expanded view of strategic control.

The processes which Figure 4.10 illustrates are as follows: the first step, naturally enough, is to gather data about the state of affairs in the environment. Since the potential amount of information in the environment is infinite, there must be some form of selection or filtering process and this selection of information is mediated via a model of the environment which specifies the parameters of interest. (The situation is analogous to the one already discussed for the feedback situation.) The processes by which this model is built up are not indicated in the diagram, but do not differ in principle from the methods used to build the models used in feedback control. It is worth pointing out that the model will be influenced to some extent in its selection of parameters by the goals of the organisation. Furthermore, in principle at least, these models can evolve in the course of time to provide better approximations of outside reality.

Once the necessary (or believed to be necessary) data has been gathered, it is used to forecast the future state of affairs. This predictive act is a vital element; its purpose is to gain enough time to allow the remainder of the functions to take place and the end result to be co-ordinated with changes outside the organisation. Precise timing may or may not be important, depending upon the nature of the organisation. A fashion business for example, must time its changes to coincide almost exactly with changes in the mood of its customers; suppliers of heavy capital plant, on the other hand, can take years to adopt a technological advance.

The diagram shows that the forecast of the future is derived by feeding the data into a model of the environment. There are some pertinent points here.

In the first place, the model used is shown as being separate from the one used to gather the data. This is to cater for practical possibilites rather than theoretical necessities; in practice, it may well be the case that two separate and different models are used for gathering information and for processing it. It is, of course, theoretically desirable that the two should be at least conformable with each other if not isomorphic one with another. However, there is no prior reason for assuming that such will be the case. If, as may happen, the models are not computer-based or not even explicitly stated, but intuitive mental models held by two or more managers, then there may well be significant differences between them.

Secondly, there is once again no indication of the way in which the model is built up originally, or subsequently modified in the light of experience. The prime reason for this omission was to avoid complexity in the diagram, but it must be admitted that it would be possible for a manager to attempt to operate without periodically updating his model (as distinct from updating his information). Taking the argument a little further, it can be seen that strategic control can easily reduce to what is effectively open-loop control; in principle, once the environmental model has been set running with the initial conditions specified by the input data, the model can continue to run without further reference to the external world. It is tempting to speculate how far organisations do actually function on these lines; although no hard evidence is available, some recent events suggest that planning procedures in some Government departments are close to an open-loop situation.

There would, of course, be nothing wrong with an open-loop situation if the model in use were sufficiently accurate to provide continuing correct forecasts. How far it is worth investing resources in improving the model is a question of some interest.

To attempt to provide a basis for answering this question, it is useful to start by hypothesising a fully determinate universe — i.e. one in which there is no quantum limit to the possible accuracy of observation and modelling. (The quantum limit can be introduced into the argument at a later stage, if required.) In order to provide a fully accurate and detailed forecast of the future it would be necessary to specify the position and momentum of every elementary particle in the universe at a given instant, together with the laws that govern their motion and interaction. This full specification is necessary if the model is to predict the *exact* course of future events, for events in distant galaxies have an effect on earth — and not necessarily an infinitesimal effect. Thus science in general, and navigation in particular, have been influenced by the study of the stars.

Once such a model had been set up, it would be necessary to find a medium on which to run it, and to supply the energy needed to drive it. At this level of detail, it becomes appropriate to talk not of a model but a replica, which puts the problem in its true perspective.

And such a replica would not necessarily be of any use. If it is to supply predictions of future states, then it must be able to compute these faster than reality achieves them, This would seem to imply that the modelling medium is capable of supporting communication at speeds greater than the speed of light.

All in all, the prospects for achieving such a model appear unpromising, to say the least. Since this is so, perhaps the tricky question of self-reference in it can be put to one side.

This line of argument indicates the difficulties likely to be encountered in pursuing predictive modelling to the ultimate. It does not lead to the conclusion that limited attempts at forecasting are of no use, if the requirement for full and absolute precision of forecasts is relaxed — or, from a slightly different point of view, if the requirement for absolute control is relaxed to one of adequate control. The problem then becomes to construct a model that will enable forecasts of acceptable accuracy to be made within a time-scale that enables use to be made of the forecast. The means by which these simpler models can be constructed lies in the redundancies and statistical properties of data gathered from the real world. Thus, to illustrate the point, it is possible in principle (except for quantum limitations) to calculate the paths and collisions of individual molecules of a gas held within a container. Amongst other things, these calculations would enable the instantaneous pressure on any part of the container to be calculated. It would be a fairly lengthy calculation — it would cover 10^{20}—10^{33} particles each with 6 degrees of freedom, but it could be done. On the other hand, the simple equation

$$pv = mRT$$

(the ideal gas law) would in all probability serve to calculate the parameters of significance in a practical problem to a satisfactory degree of accuracy, and provide the answers much more rapidly.

In practical terms, in the context of organisation and strategic control, the problem is where to strike the balance between a fully detailed but cumbersome and slow model, and an approximate but rapid one. It would appear that theoretically there is no absolute answer to this problem, but that the answer is contingent upon the nature of the organisation that is attempting to use the model. In particular, it is related to what might be termed the 'reaction time' of the organisation in question, i.e. the time which it takes for an undertaking to make significant changes to its product. The forecast needs to cover at least a sufficient period ahead to enable the organisation to adjust itself to predicted change. Thus, there is little point producing a forecast for the next six months if it takes five years for the organisation to change its operations. Equally, there is little point in a forecast extending ahead more than say 10 or so 'reaction times', it is an unnecessary use of resources to plan ahead much further than this, because the organisation will have ample time to adjust to changes beyond this time-scale. Furthermore, it is in the nature of forecasting that the further ahead the forecast is made for, the less precise and reliable it becomes.

As a consequence of this approach, it can be concluded that the need for, and nature of, strategic control will be a function of the nature of the organisation in question, with size of organisation being a very relevant variable. A small organisation which can adapt very rapidly to change in the environment will have a limited need for this type of activity (indeed it may be possible for it to survive for an appreciable period without it). A large organisation will require much more sophisticated forecasting techniques.

However, once a forecast has been made, the next step is to evaluate the likely effect on the organisation, whether adverse or favourable. (In many ways, this is analogous to the process of comparing actual results with goals in a feedback loop.) As shown in Figure 4.20, a necessary input at this stage is some form of organisation model, relating in particular to objectives and long-term plans. These are compared with the predicted future state of the environment, and discrepancies sought. Mismatches between the two indicate a need for the organisation to undertake some action. In principle, there is no reason why the organisation should not attempt to rectify a mismatch by changing the future course of the environment. However, such a course of action falls outside the scope of the present discussion, and falls more naturally into the category of external regulation. Discussion of such a course of action will therefore be postponed.

The basic process remaining is to adapt the organisation for the expected changes in input. There are two distinct aspects to such a process of adaptation. The first is what might be described as parameter adjustment, i.e. setting new goals, more appropriate for the future as forseen. This is the type of process envisaged by Jankowicz, as has been discussed above. In principle it accomplishes nothing that could not be achieved through feedback, with the proviso that strategic control of this type allows a faster response — even an anticipatory response. In a competitive environment, factors of speed of response can be important.

The second aspect of strategic control is a modification to the fabric of the organisation to enable it to cope better with the foreseen environment. These modifications may be either to the productive base of the organisation (new products, new plant, etc), to the managerial superstructure built on the base (a reorganisation) or a combination of both.

Such modifications to an organisation (particularly modifications to the productive base) can only be achieved at a price. Resources need to be applied, and the amount of modification possible will be determined by the amount of resources available. In the case of a business, the amount of resources available is determined by the profitability of the enterprise. (Not necessarily directly, due to the fact that money can be borrowed, but the amount that can be borrowed bears a relationship to ability to repay, and hence to profitability.) This need for modification to the business explains the need for profit, and also suggests that profit needs to be higher in an uncertain environment. It also suggests that the profits that a firm requires can be calculated.

It is the capability of undertaking this type of activity that distinguishes organisations from entities in the natural kingdom. It is equivalent to growing extra limbs or reshaping the neural pathways of the brain. The fact that such adaptations are possible increases the potential variety that organisations can cope with; the fact that such adaptations occur indicates that organisations function in an environment of a higher order of complexity than that of the physical universe that ordinary organisms exist in.

However, the laws of cybernetics still apply to this situation, in particular Ashby's Law of Requisite Variety. As has been pointed out, Ashby's law covers the basic mechanism of strategic control — indeed, it was first set-out in that form, with feedback control as a subsidiary modification. Thus, the amount of regulation that can be achieved through strategic control is limited by the channel capacity of the control path.

Limitations on channel capacity are usually thought of as largely physical problems, associated with the rate at which information can be passed through a communication path. Whilst such factors can (and in many instances, undoubtedly do) limit the capacity of a particular channel, they are not the only possible source of limitations in channel capacity. The other source of restriction on channel capacity, which appears not to have been discussed in the literature, is what might be termed modelling capacity.

The fundamental concept that this term is intended to convey is that control over any situation is achieved by processing information through a model (see Ashby and Conant, 1970) and the limitation on channel capacity may well derive from the rate at which the model can process information rather than from the rate at which information can be transmitted to and from the model. (Indeed, it can be argued that the capacities required for information transmission should be calculated from the processing rate of the model.)

This limitation on channel capacity arising from a model can be seen most readily in the case of a digital model. We are accustomed now to the idea that a given computation takes a certain amount of time. Thus, if the model we are using has three variables, x, y, z, then the time required to compute the outcome depends upon the functions used, i.e.

$$R = 3x + 3y + 3z \, .$$

is quicker than

$$R = (\sin x + \cos^2 y + \tan z/2)\, (3x + 3y + 3z)$$

Thus (assuming that R is to the same accuracy in both cases) the rate at which R can be computed (i.e. the modelling capacity) depends upon the complexity of the model. It will also depend upon the number of inputs (and outputs) required, e.g.

$$R = \sum_{1}^{5} x_n$$

is quicker than

$$R = \sum_{1}^{100} x_n$$

The same limitation on processing capacity is also found in analogue models, though it is expressed in different ways, usually in terms such as transient response.

However, the most important point is that it is the model (or models) in use that form the essential limit upon channel capacity. The model in use will determine what input and output are required and to what accuracy. It will also determine the number and nature of calculations required to derive the output from the input. These factors, together with the speed of the computer used, will determine the maximum channel capacity available.

As shown in Figure 4.10, there are essentially at least two models present in the strategic control loop, one a model of the environment, used for prediction, the other a model of the organisation used to determine the changes required to meet the foreseen future. Apart from the fact that they are models of different things, there are important differences between the essential requirements for these two models.

The nature of the environment model is such that it is essentially variety - reducing, in that it seeks to predict the course of a limited number of key variables from information taken from a variety of sources. Furthermore, it can, in principle, be a black box model; as long as it produces usable results, its internal workings are not necessarily of great relevance.

By way of contrast, the organisational model has the opposite characteristics. It is variety-generating, in that the input from the environment model is used to generate the required changes throughout the organisation. Furthermore, it cannot be a black box model; in order to generate the required modifications, the model must show at least some of the internal structure and connectivity of the organisation. The range of possible modification to an organisation is then the permutation of the internal inputs and outputs, coupled with the changes that can be wrought to each component function through investment, and the essence of the planning process is to extract the optimum from this range of possibilities. Naturally, the more detailed is the model used, (the more internal structure is shown) the more numerous are the possible courses of action.

The foregoing outlines the essential cybernetic requirements for the models used in strategic control. (It may of course be the case that in practice these requirements are exceeded.) It does not necessarily follow that the functions are readily identified with the work of any particular individual or group of individuals. The models discussed are not necessarily embodied in computer programs — or indeed even set out formally at all. They may be distributed across the members of the organisation, particularly the management of the organisation. (Indeed, such informal models will always exist, even where formal computer models have been constructed.) Nor is it at all likely that such informal, distributed, models will all be in total agreement one with another.

What is more, changes to an organisation rarely affect just one isolated section of it. Most changes affect considerable sections of an organisation, some involve all of it. Planning, therefore, generally involves large sections of management, acting horizontally across the hierarchy. This may take the form of committee work, or the setting up of an informal network of communication (the "informal organisation" of management literature) or a combination of these. It is a considerably different mode of organisational activity from the traditional bureaucratic hierarchy, which is most likely to be evident during periods of organisational change. This view is endorsed by the work of Burns and Stalker (1961), who observed that organisational innovation typically brought forth new forms of managerial behaviour. It would seem that the cybernetic explanation of this phenomenon is that innovation requires an interactive, unified, approach, across the whole organisation; it is likely, furthermore, that management will be heavily involved in restructuring their models of the organisation during such a period of change.

A further consequence that can be anticipated to stem from strategic control is a cycle from (in management jargon) centralised to de-centralised and back to centralised forms of organisation. When the organisation is making major adjustments to fit a new environment, a relatively high degree of central co-ordination will be required, and hence a centralised form of management will be appropriate. As the organisation settles down in its new role, it is appropriate to allow the component parts of the organisation some freedom to "fine-tune" their operations (by a process of Ashbeam adaptation) and a more decentralised form will be appropriate.

Thus there is at least some cybernetic justification for the well-known business phenomenon of a cycle from centralisation to decentralisation. The explanation does not necessarily cover all instances of the phenomenon — firms may engage in the cycle for quite other reasons, more to do with personalities and politics — but it does cover some instances.

Overall, the process of strategic control can be clearly differentiated from administrative management. The purpose of strategic control is to design (or redesign) the enterprise to provide the desired results. The purpose of administrative management is to operate the organisation to actually achieve the desired results. Although there are quite separate functions, this is frequently not recognised in organisation structure, and very often both are carried out by the same individuals. The situation may be further complicated by the fact that the original design for the organisation may not have been totally correct, and operating management need to make some adjustments. However, although lines may be blurred in practice, the main features of both are clear.

4.3 External Regulation

4.3.1 Output environment

As has been mentioned previously, organisations seek to influence the environment as well as their own internal affairs, and this influence can be divided into two broad categories. The first, which will be discussed here, is the category of influencing or controlling the output environment, i.e. the market for the organisations goods and services.

This is not generally recognised as a specifically managerial activity, although it is widely acknowledged as a function of organisations, particularly of business organisations. (Its most obvious manifestation is in the form of advertising and kindred activities.) The reason for including it here is that it is obviously a form of control activity undertaken by organisations, and therefore of cybernetic relevance.

Having established that, there is not a great deal more that requires to be said. The general methods used are well known — advertising, pricing, public relations — based on a comparatively simple model of economic behaviour. The most interesting question surrounding these operations is to what extent they can hope to be effective, i.e. to what extent can an organisation control its market? Cybernetics would suggest that the answer is only, to a very limited degree, an answer supported by experience.

It can, however, be pointed out that there appear to be special difficulties in managing this type of operation. Mention has already been made of the fact that management operates on models of the outside world. But in the marketing and advertising sphere there are so many variables — all of them probably interacting, many of them perhaps unknown — that it is very difficult indeed to test a model against reality in any systematic way. Almost anything may work, and equally the most carefully thought-out scheme may fail. To put it another way, the market-place is a very black box indeed. This may offer an insight into why originality and creativity are more often found in the sphere of sales than of manufacturing.

4.3.2 Input environment

As well as attempting to regulate their markets, organisations attempt to influence the society and culture within which they operate. This is at least in part because of the profound effect that attitudes, customs and laws have upon the operations of an organisation and the market for its products. For example, Factories Acts have effects upon methods of production, manning levels, shift work and the like; various Road Traffic Acts have a great influence on the design of motor vehicles; taxation can have more effect on price levels than any other factor, particularly for tobacco and alcohol products.

Given that such factors influence the operation of an organisation, it can be to the organisation's advantage to have as much control as possible over them.

The basic functions needed have already been outlined in Figure 4.10 when discussing strategic control. A model of the environment is used to predict what is likely to happen, and the consequences for the organisation evaluated.

However, instead of using the result to control the organisation, it is used to influence the environment.

However, in order to be able to do this, there is one important modification needed, which concerns the nature of the model of the environment. For strategic control, a simple black box model of the environment was all that was necessary; but to control the input environment, this will not suffice. It is out of the question for the organisation to control the environment directly; one reason is the relative resources of each, and another is that the organisation does not have access to the inputs of the environment. Thus in order to exert some regulation on its environment, the organisation needs to be able to locate the centres of power in the environment and gain access to them.

It is a feature of a black box model that it does not identify the centres of power, or indeed anything beyond a simple input-output relationship. A model with more structure is needed — a "grey box", as it were, with at least some of the internal nodes accessible.

Many organisations employ people whose major contribution to the enterprise can be construed as knowledge of how the environment is structured, and who can gain effective access to at least some of the power centres. Such people are usually found at very senior levels within an organisation, and may often contribute little or nothing to the day-to-day operation of the enterprise. Yet, as can be seen, their contribution can be vital.

In addition to this type of direct access to centres of influence, organisations often form into groups for the basically political purpose of forming a pressure group to represent their interests. Examples of such groups are easy to find, ranging from Guilds through Chambers of Commerce to the CBI.

4.4 Summary

The presentation of the model of organisation is now complete, and this is a convenient point to summarise what has been said.

The model has been based on cybernetic principles of control. It follows the general approach used by other commentators such as George (1970, 1974) and Jankowicz (1973) but develops it in greater detail. In doing this, some extra concepts have been introduced, the main ones being:

1) Not all elements in a manager's goal vector are derived from the goals of the organisation. It is impossible to employ a fraction of a person, and thus it is inevitable that personal aspirations and targets will enter into executive decisions. This has far-reaching consequences, and is perhaps the most fundamental distinction between organisational control systems and others.

2) Elements within the goal-vector may be incompatible one with another. This can arise either from conflict between the purely organisational goals or from conflict between organisational goals and personal ones. These difficulties can usually be resolved because of the inherent latitude in the specification of the objectives.

3) There need be no logical relation between elements in the goal-vector. This occurs when a number of essentially disparate functions are brought together under the management of one individual.

4) Control is exercised through the use of models, which are the result of experience. A wide variety of models can serve as viable controllers of the same operation.

5) For many purposes, a black box model can provide perfectly adequate control (though in practice more complex forms may be used). The exception to this rule is planning activity, where a model with at least some internal structure is necessary.

As the approach has been set out, various comments have been made to show how it can account for the reported features of organisations. Since these are rather scattered in the text, it seems appropriate to summarise the main points here.

The organisation has been characterised as an essentially dynamic entity, with the need for management arising from the perturbations that occur. The typical hierarchical structure found in institutions was explained as an expression of a system of nested control loops. Some initial reasons for this formulation were given, but these are only a part of the story. The fundamental reason for nested loops arises from the limited information-handling capacities of people, a topic which is explored more fully later on in this book.

The wide variety of structures found in practice can be seen as arising firstly from a somewhat arbitrary division of the total operation into parts, followed by an equally arbitrary set of decisions as to what control loops should be installed. The grouping together of these control loops into tasks for individual managers is another area where differences between organisations can arise.

A definite distinction has been drawn between administrative and planning activities, based on the type of model required for each. This distinction is carried further when plans are put into action, which calls for discussion and co-operation horizontally across the enterprise rather than the vertical communication that characterises administrative management.

The importance of the human element has also been discussed. This was introduced first via the personal elements in a goal-vector, which leads to the well-reported phenomenon of conflict. The point can be made that there is nothing surprising or distinctive about such conflict; it is a very general feature of systems with multiple goals rather than a result of the presence of people. Conflict is frequently found among organisational objectives; the addition of personal elements generalises this across a wider set of elements.

The further implications of the human element were explored in terms of "unofficial" paths of communication. This is perhaps a feature that is specifically attributable to the presence of people. It was shown how these extra paths would result in behaviour that could be interpreted variously as "informal organisations", "working groups" and the like. Attention was also drawn to the effect of imperfect organisation design on these processes.

The relation of the organisation to its environment has also been touched on, though not in great depth, and again it has shown how this can be accounted for in terms of cybernetic mechanisms.

It can be seen therefore that the functions of organisation and management can be interpreted in terms of well-known theoretical concepts. Due attention must be paid to the special features that are involved, but there is no difficulty of principle involved. This in turn opens the way for a rational and scientifically-based understanding of organisations and their effectiveness.

5

Some Consequences of the Model

5.1 Introduction

Having developed a model that accounts for many of the features of organisations, there are two further steps to be taken. The first is to test it against reality by applying it to practical situations and practical problems. The other is to develop the theoretical implications, to make predictions of what should be the case.

The remainder of this book is devoted to this exercise. With such a wide field to choose from, it cannot be pretended that the treatment is in any way exhaustive. It is much better regarded as a preliminary indication of some of the possiblities.

To start with, the results of a live study are reported. It is hoped that this will be of some interest in its own right, but equally important is the way that it shows how meaningful research can be carried out using the model. It suggests strongly that quantitative measures can be applied to many of the problems of organisation design.

Following on from this, some theoretical consequences of one aspect of the model are examined. These, it is believed, provide some useful insights into the complexities of the management process.

5.2 A Practical Study

The analysis and experiment reported here arose from a practical requirement in a major British company. For commercial reasons, not all of the work undertaken can be reported here, particularly those aspects bearing on profitability.

The analysis formed part of a larger study of the work of Sales Managers, each of whom was totally responsible for the operation of a number of retail outlets. (Various support staff were available to assist in staff capacities, but the Sales Manager was the clear focus of responsibility.) The chart in Figure 5.1 shows the organisation.

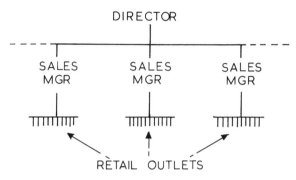

FIGURE 5.1.   An organisation chart of the Retail Sales Management operation.

Amongst other things, the larger study revealed that the number of outlets allocated to each Sales Manager varied widely, between approximate limits of 20 to 40, i.e. a 2-1 ratio. Graph 5.1 shows the actual distribution of outlet allocations. There appeared to be little if any scientific rationale behind these varying numbers, and the question arose as to what was the optimum number of outlets per Sales Manager.

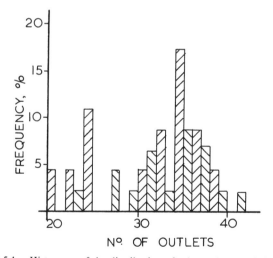

GRAPH 5.1   Histogram of the distribution of sales outlets per Sales Manager.

Established methods of ascertaining work load were examined (i.e. variants of Time Study procedures) but none seemed appropriate to this type of work. However, the study had already established that approximately 95%

of the Sales Manager's time was devoted to the "administrative" or "feedback" aspect of management, as defined previously. Therefore it was decided to investigate the use of the concept of channel capacity as a means of resolving the problem.

(In passing, it can be noted that the problem of how many outlets a Sales Manager can control can be answered from two different viewpoints, that of the Sales Manager — How many can he cope with in a working day? — and that of the Company — What is the best allocation for optimum profit, including the cost of Sales Managers? The two answers are not necessarily the same. The work reported here is concerned fundamentally with the former of the two approaches.)

Channel capacity was applied to the problem as follows. Hick (1952) showed that the human operator can be regarded as a channel of limited capacity. Typical behaviour at various rates of information flow is shown in Graph 5.2. At low rates of input, the human operator functions as a virtually perfect information channel, information out equalling information in. At higher rates, performance falls off slightly and there is some loss of information. This fall-off is approximately linear until the limiting channel capacity is approached. This maximum channel capacity was of the order of 7 bits/second. However, this figure was not maintained as the input rate was increased; channel capacity fell off quite markedly as the input rate was increased beyond the point of maximum capacity.

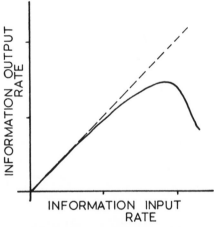

GRAPH 5.2.   The general form of the behaviour of a human operator viewed as a control channel.

Other work has confirmed this general shape of curve, and shown it to be a typical property of information processing systems. A good summary of the evidence can be found in Miller (1962).

The Sales Manager can be considered as a control channel over his retail outlets. Furthermore, since each outlet is independent of the others (i.e. functions in parallel with them, not in series) the information input to the Sales Manager is a linear function of the number of outlets he controls. (This is true on average, if outlets are assigned at random from a statistically homogeneous population; the effects of such statistical variation are considered below.) Furthermore, the channel capacity required to control them is also a linear function of the number of outlets.

If the Sales Manager channel capacity follows the form of Graph 5.2, the control he exerts over his outlets can be expected to vary in the general way sketched in Graph 5.3 as the number of outlets varies.

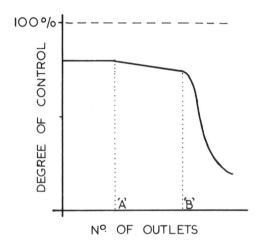

GRAPH 5.3.   The predicted relationship between number of outlets and degree of control.

Even at a low workload, control would not be perfect (i.e. would be less than 100%) because the manager is operating in a feedback mode, which leads to residual error in the controlled output. Thus, Graph 5.3 starts at less than 100% control. However, at a low workload, the Sales Manager can maintain control at this level as his number of outlets is increased. Eventually, however, at A outlets on the graph, his channel capacity starts to fall below the capacity required, and overall control starts to decline. This decline will be progressive until, at about point B on the graph his maximum channel capacity is reached. Thereafter, control declines rapidly, but probably not to zero.

Clearly, a graph like this would enable the optimum allocation of outlets to be determined, by determining where the points A and B fall. The optimum figure is a little beyond the point A, sufficient load to set a challenge, but not approaching point B, the point of overload. There is little point in operating outside this range; workloads lower than A produce no benefits, and beyond point B there is little point in having a Sales Manager, for he is almost totally ineffective.

This was the basic theory which it was decided to use to attempt to answer the question as to how many outlets a manager should control. Clearly, such a simple theory could not be expected to account fully for all the factors likely to be encountered in practice. For example, the theory assumes that all outlets are identical, which is certainly not the case in practice. The average sample size (i.e. outlets per Sales Manager) was of the order of 35, which, while a reasonably reliable sample, was not guaranteed to even out all inconsistencies. Some further variables not accounted for are as follows:

1) The ability and experience of the Sales Manager.

2) The overall geography of his area (e.g. compact or dispersed).

3) Level of support staffing.

4) Quality of staff in the outlets themselves.

5) Local trading conditions.

Each of which could be expected to have some effect. Thus, it was to be expected that any results would show a considerable degree of scatter. Indeed, it was possible that the scatter would be sufficient to completely mask any effects due to workload.

The theory also left unresolved the question of how degree of control was to be measured. The basic definition of control can be taken (vide Ashby, 1956) as

$$C = (1 - V_0/V_1) \times 100\%$$

where

C = degree of control
V_0 = range of controlled output
V_1 = range of input

The actual controlled output of an outlet is a vector with many components, including such items as staff morale, public relations, etc. The actual input is of similar complexity. However, it was decided that a satisfactory estimate of control could be obtained from examining the relationship between the takings of an outlet and its overall profitability. Quite apart from the fact that detailed data on these variables was available, it was apparent that they formed the key objective of most of the Sales Manager's work.

The calculations used to arrive at V_0 and V_1 for an individual outlet can be

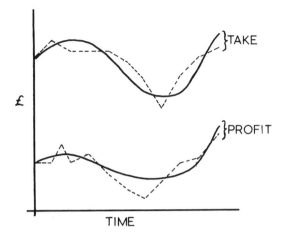

GRAPH 5.4   Schematic graphs to illustrate the principles of the calculation of the control index.

illustrated by Graph 5.4. Figures were available for the forecast and actual values of takings and profit, which typically showed the general pattern of Graph 5.4, with random fluctuations imposed on a seasonal trend. In principle, V_0 was the variability of the profit figure, whilst V_1 was the variability of the take. Certain corrections were applied to this basic scheme, arrived at as follows: If the takings were absolutely constant throughout the year, then the variability of the profit could be used as an index of control (periodic charges, e.g. rates, electricity, were spread evenly throughout the year by the accounting procedures in use). However, since the take is not constant, corrections need to be made.

Firstly, the forecast shows that the volume of trade is expected to vary through the year, and operating methods need to be adjusted to cope with this variation, e.g. extra staff need to be taken on, more stocks purchased, and so on. The greater the expected variation (i.e. the more markedly seasonal the trade) the greater these adjustments need to be, and the more critical is the timing of them. Individual outlets varied widely in the seasonality of their trade; for some it was no more than 5% of the annual average, whilst at the other extreme some outlets approached 100% of the annual average. The seasonality of the trade, as indicated by the forecast, was thus a factor that needed to be accounted for.

The other factor considered was the variability of the actual take against the forecast. If the actual takings differ by a constant amount from the forecast throughout the year, then the difficulty of controlling the outlet does not increase. If the actual differs from the forecast by a variable amount, then the difficulty increases, in proportion to the variability of the difference.

Furthermore, the profit figure needs to be adjusted to take account of the variation in takings. To this end, actual profit was expressed as a percentage of actual take, and compared with the forecast percentage profit (obtained from forecast) in this way made allowance for the fact that expenses do not vary in strict proportion to trade, due to fixed expense elements. This method of correcting for the fixed element is not absolutely accurate, but is approximately true when working well above the break-even point, as was generally the case. What is more, any inaccuracies introduced apply consistently across all Sales Managers and thus should not affect the final result.

With these corrections, V_0 and V_1 became as follows
V_0 = standard deviation of (Pa - Pf) where

Pa = actual profit%
Pf = forecast profit%

and

$$V_1 = S \times F$$

where S = standard deviation of the forecast, expressed as a percentage of the forecast average take

F = standard deviation of $((Ta \_ Tf)/Tf) \times 100\%$ where
Ta = actual takings
Tf = forecast takings

All measures were computed over one financial year for each outlet. It is worth noting that the measures used were all pure numbers, and that since variances rather than averages were used, any systematic errors in the forecast would be cancelled out and not affect the data.

Thus a control index could be calculated for each outlet over a year, using the formula

$$C = (1 - V_0/V_1) \times 100\%$$

To calculate the index for a Sales Manager, the mean value for all outlets under his control was calculated, with the proviso that the outlet must have been trading continuously under his control for at least 18 months. This proviso excluded outlets in the following categories.

1) Outlets transferred recently from another Sales Manager

2) New outlets recently acquired.

3) Outlets temporarily closed.

It was felt that such outlets should be excluded because the Sales Manager would not be fully familiar with its operation. However, outlets excluded on this basis were included back in when arriving at the number of outlets under his control. Only a small number of outlets — rarely more than 2 or 3 per Sales Manager — were excluded in this way.

The method thus developed was applied to a pilot sample of 8 Sales Managers. The sample was selected using the following criteria.

1) Sales Managers should be from the same geographical area, working under the same Director, to hold constant as many extraneous variables as possible.

2) The sample should include as wide a cross-section as possible of number of outlets per Sales Manager.

Data were obtained by manual extraction of figures from 4-weekly $P + L$ accounts for each outlet, and processed with the aid of an HP 65 programmable calculator. The results obtained were as follows.

Number of Outlets	Control Index
23	95.8%
26	94.2%
30	92.8%
31	94.3%
34	93.2%
34	91.1%
38	91.7%
38	90.4%

These results are also shown in Graph 5.5

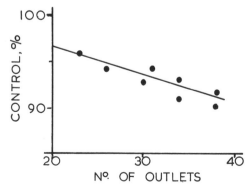

GRAPH 5.5   Results from the pilot study.

Fitting a straight line to the results yields the equation

$$y = 102.43 - .30x$$

with a correlation of 0.88.

Thus the pilot sample confirmed the basic theory with a high degree of success, with all results falling in the region of declining performance but with no evidence that the point of breakdown was being reached. Nor was there any indication of where the point of overload might lie. A lower limit was found by extrapolating the results to cut the y = 100 line — this gives a lower limit on A. The value of this limit was given by

$$100 = 102.43 - .30x$$
Hence
$$A = 8.1$$

The results were sufficiently encouraging to extend the method to a full-scale survey. The data extraction and analysis were performed by IBM370 computer, and thanks are due to Mr C. Holmes and Mr J. Perry who undertook the necessary programming.

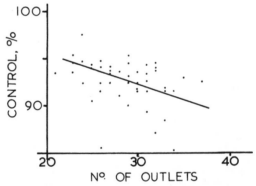

GRAPH 5.6   Results from the full-scale study.

These results are plotted in Graph 5.6. It is apparent that there is a greater degree of scatter than in the pilot study, which is to be expected due to the inclusion of variables which were minimised in the sample, such as trading conditions, support staffing, director's influence amongst others. Further-more, there are some results which do not fall in with the main trend, notably at low numbers of outlets, and a group lying above the apparent main trend. Inspection of this latter group showed that they all came from one geographical region, and that all Sales Managers from that region fell into that group. Thus it could reasonably be inferred that there were special, unidentified factors in operation for that group which can therefore be excluded from the main analysis.

Fitting a line to the remaining results yields a correlation of 0.338, which is significant at the p = .01 level.

Averaging the results at each number of outlets — thus averaging out the effects of the random variables mentioned above — yields the results shown in Graph 5.7. Fitting a straight line to these yields a correlation of 0.851, again significant at the p = .01 level.

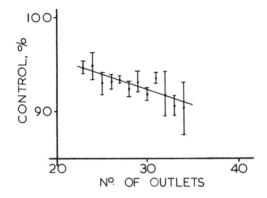

GRAPH 5.7   Smoothed results from the full-scale study.

These values of the correlation coefficient show that the data provide strong experimental support for the original proposition. From a practical point of view, the results as they stand do not answer the basic question with any precision, in that the location of point A is still in some doubt. It is clearly outside the range of the main body of results, and the few results for low numbers of outlets are not sufficient to locate A with any precision. Some extrapolation of the results is necessary, and this can most conveniently be done via a means such as Graph 5.8. This is a reconstruction of the information input vs. information output graph of Graph 5.2. Each axis is plotted in terms of number of outlets, and the diagonal line at 45° therefore represents perfect information transmission. The points on the graph are obtained from the data by multiplying each number of outlets by its associated control index, which in turn is a measure of channel capacity as a fraction of channel capacity required. Drawing a line through these points to intersect with the diagonal locates the point A, which turns out to be at 19.6 outlets per Sales Manager.

The results obtained are important at a number of levels. At the lowest level, they provide a definite answer to the original practical problem. This particular answer applies strictly to the environment in which it was obtained

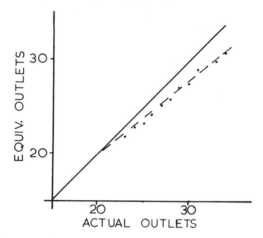

GRAPH 5.8.   Reconstruction of an information transmission relationship from the experimental results.

— the value for *A* could well differ for a different organisation in the same branch of retailing, would almost certainly differ for a different type of retail trade. However, it is clear that the method is sufficiently general to be applied to similar problems with every hope of success.

These similar problems need not be confined to retail sales management. The basic philosophy of the method could be applied to problems such as the optimum size of classes in schools, or the desirable manning level in the Police Force, as well as a variety of industrial situations.

An aspect of this work which should not be overlooked is that it could be developed to form a quantitative basis for the assessment of managerial perfomance (or at least the administrative aspect of performance). The results of an individual manager could be measured against the average value at any particular workload. Whilst this would provide only a comparative measure against his colleagues, rather than an absolute value, it would have the merit of avoiding entirely any subjective element in assessment.

At another level, the results provide confirmation of at least one aspect of the model of organisation that has been propounded. As far as can be ascertained, this is the first report of any direct evidence supporting the general feedback model for organisations.

At a final, and most important level, the results demonstrate that it is possible to undertake meaningful quantitative research in the field of organisation structure and design. Given the enormous and growing importance of organisations in everyday life, the ability to subject them to scientific scrutiny cannot be overstated. The work reported here forms a first step towards such an end.

5.3 Some Theoretical Aspects

Having developed and tested a model for organisation, it is of interest to examine some of its features and the implied consequences. The features which it is proposed to examine here have as a common theme various aspects of the modelling process of which mention has been made. It is a process which appears to have been largely taken for granted in much of what cybernetic work has been applied to organisation, yet it is an issue of central concern. The particular aspects of it which will be discussed here are:

1) The nature of managerial feedback controls.
2) Consistency of models among managers.
3) The nature of the modelling process.
4) Speed of data processing.

5.3.1 The nature of managerial feedback loops

Given that a significant part of management activity can be characterised as feedback control, it is of interest to examine the nature of this process in a little detail. It should be emphasised that what follows is a gross simplification of reality. In practice, managerial feedback is essentially a sampled data system, controlling non-linear, even non-analytic systems with many interacting variables.

However, managers' ideas of feedback do not normally encompass this degree of complexity, and many management information systems are designed on the basis of an extremely simple notion of feedback. There is therefore some element of validity, as well as the merit of simplicity, in examining the simplest possible model of feedback.

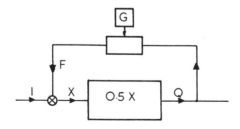

FIGURE 5.2.   A simple model of feedback activity.

As an initial example, consider a control system such as that shown in Figure 5.2. Suppose the equations are:

1) $x_t = I_t - F_t$
2) $Q_t = 0.5\, x_t$
3) $F_t = Q_{t-1} - G_{t-1}$

i.e. a simple negative feedback situation. If I is set to 10, G to 5, and the system set off, then the trajectory shown in Table 5.1 results.

TABLE 5.1

t	I	F	x	Q	G
0	10	-5	15	7.5	5
1	10	+2.5	7.5	3.75	5
2	10	-1.25	-1.25	5.62	5
3	10	+.62	9.38	4.69	5
4	10	-.31	10.31	5.16	5
5	10	+.16	9.84	4.92	5
6	10	-0.8	10.08	5.04	5
7	10	+.04	9.96	4.98	5
8	10	-.02	10.02	5.0	5
9	10	+.01	9.99	4.99	5
10	10	-.01	10.01	5.00	5

(See also Graph 5.9.)

The system finally stabilises to the goal (5), but only after a series of fluctuations. The "uncontrolled" system would have reached the goal immediately. When the input varies, a trajectory such as Table 5.2 is obtained.

TABLE 5.2

t	I	F	x	Q	G
11	10	0	10	5	5
12	11	0	11	5.5	5
13	12	+.5	11.5	5.75	5
14	13	.75	12.25	6.12	5
15	14	1.12	12.88	6.44	5
16	15	1.44	13.56	6.78	5
17	14	1.78	12.22	6.11	5
18	13	1.11	11.89	5.99	5
19	12	.99	11.01	5.50	5
20	11	.50	10.50	5.25	5
21	10	.13	9.75	4.87	5
22	10	.25	10.13	5.06	5
23	10	.06	9.94	4.97	5
24	10	.03	10.03	5.01	5
25	10	.01	9.99	5.00	5

(See also Graph 5.10.)

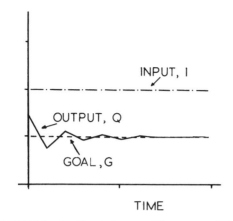

GRAPH 5.9.   Starting trajectory of the system of Figure 5.2.

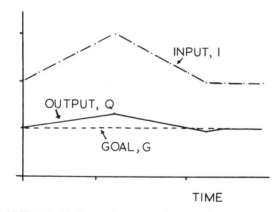

GRAPH 5.10   Trajectory for a ramp input to the system of Figure 5.2.

This is not a very impressive performance. The control index

$$C = (1 - V_0/V_1) \times 100\%$$
$$\text{is } C = (1 - (6.78 - 4.87)/2.50) \times 100\%$$
$$= 4.6\%$$

obtained as follows. V_0 is the actual range in the controlled output, from a maximum value of 6.78 to a minimum of 4.87. V_1 is the range in the output that would have occurred without the intervention of the control system, i.e. the range in the input multiplied by the forward transfer function, in this case $(15 - 10) \times 0.5$.

A different form of input variation gives a different type of trajectory, as shown in Table 5.3.

TABLE 5.3

t	I	F	x	Q	G
26	10	0	10	5	5
27	10	0	10	5	5
28	15	0	15	7.5	5
29	15	2.5	12.50	6.25	5
30	15	1.25	13.75	6.87	5
31	15	1.87	13.13	6.57	5
32	15	1.57	13.43	6.71	5
33	15	1.71	13.29	6.64	5
34	15	1.64	13.36	6.68	5
35	15	1.68	13.32	6.66	5
36	15	1.66	13.34	6.67	5
37	15	1.67	13.33	6.66	5
38	15	1.66	13.34	6.67	5
39	15	1.67	13.33	6.66	5
40	15	1.66	13.33	6.67	5

(See also Graph 5.11.)

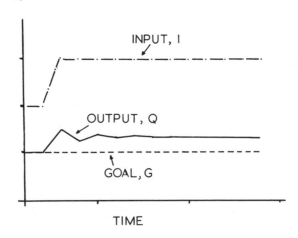

GRAPH 5.11. Trajectory for a step input to the system of Figure 5.2.

A feature to note here, apart from the long settling time, is that the final equilibrium reached is not at the goal, but at a point intermediate between it and the new uncontrolled output level. The same effect can be observed if the goal is varied rather than the input, as shown below in Table 5.4.

TABLE 5.4

t	I	F	x	Q	G
41	10	0	10	5	5
42	10	0	10	5	3
43	10	2	8	4	3
44	10	1	9	4.5	3
45	10	1.5	8.5	4.25	3
46	10	1.35	8.75	4.37	3
47	10	1.37	8.63	4.31	3
48	10	1.31	8.69	4.34	3
49	10	1.34	8.66	4.33	3
50	10	1.33	8.67	4.33	3
51	10	1.33	8.67	4.33	3

(See also Graph 5.12.)

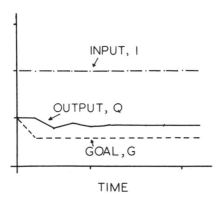

GRAPH 5.12   Trajectory for a change of goal in the system of Figure 5.2.

Here again, there is a marked discrepancy between the goal of the system and its final equilibrial value. (In passing, it can be pointed out that this serves to reinforce the arguments about the difficulty of inferring the goals of taciturn systems (Pask, 1969).)

In total, the alleged feedback control system of Figure 5.2 does not perform too well. It is of interest chiefly because much of the literature about managerial feedback (e.g. Brown (1971), Humble (1968), Donald (1967)) appears to take the view that there is no more to feedback than is contained in the diagram and its associated equations.

There was much discussion in the foregoing section of the need for models for control. Let us therefore introduce a model of the controlled system into the control loop. This can be done simply by modifying one of the equations, so that

4) $x_t = I_t - 2F_t$
5) $Q_t = 0.5\ x_t$
6) $F_t = Q_{t-1} - G_{t-1}$

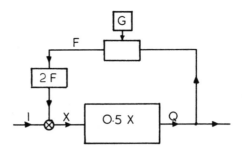

FIGURE 5.3   Showing the introduction of a model into the feedback loop.

The 'model' here corresponds to the factor $2F_t$ in equation 4–(2 = 1/Transfer function = 1/0.5). This modification does not greatly improve performance, as can be seen in the following trajectories in Table 5.5.

TABLE 5.5

t	I	F	x	Q	G
0	10	0	10	5	5
1	10	0	10	5	5
2	10	0	10	5	5
3	11	0	11	5.5	5
4	12	.5	11	5.5	5
5	13	.5	12	6.0	5
6	14	1.0	12	6.0	5
7	15	1.0	13	6.5	5
8	14	1.5	11	5.5	5
9	13	.5	12	6.0	5
10	12	1.0	10	5.0	5
11	11	0	11	5.5	5
12	10	.5	9	4.5	5
13	10	.5	11	5.5	5
14	10	.5	9	4.5	5

(See also Graph 5.13)

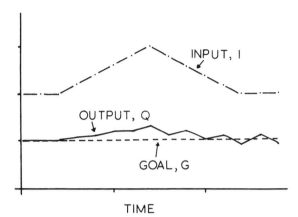

TIME

GRAPH 5.13   Starting and ramp input trajectories for the system of Figure 5.3.

For the same input function as before, the control index becomes

$$C = (1 - (6.5 - 4.5)/2.5)$$
$$= 20\%$$

Although it is a four-fold improvement on the previous situation, this must be balanced against the fact that the system has now gone into permanent oscillation. (Some readers may have noticed that originally the system could tend to oscillate in this way when switched on if the initial conditions were unfavourable.)

This system is not much better at maintaining the goal when the input undergoes a step change, as is shown in Table 5.6.

TABLE 5.6

t	I	F	x	Q	G
15	10	0	10	5	5
16	10	0	10	5	5
17	15	0	15	7.5	5
18	15	2.5	10	5	5
19	15	0	15	7.5	5
20	15	2.5	10	5	5

(See also Graph 5.14.)

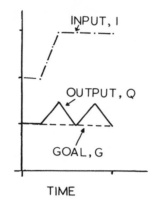

GRAPH 5.14   Trajectory for a step input to the system of Figure 5.3.

The oscillatory behaviour is still present, and, a point of some importance, the mean of the oscillations is not at the goal. The same is true when the goal of the systems is changed, as is shown in the trajectory in Table 5.7.

TABLE 5.7

t	I	F	x	Q	G
21	10	0	10	5	5
22	10	0	10	5	5
23	10	0	10	5	3
24	10	2	6	3	3
25	10	0	10	5	3
26	10	2	6	3	3
27	10	0	10	5	3

(See also Graph 5.15.)

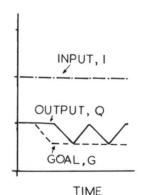

GRAPH 5.15.   Trajectory for a change in goal in the system of Figure 5.3.

The behaviour is still oscillatory, and the mean of the oscillations is still not at the goal. The control system is still not satisfactory, in spite of the fact that a perfect model has been built into it.

The initial response to the problem of oscillation is to include some damping in the control loop. However, as has been pointed out, the mean of the oscillations is not at the goal, and thus damping would not entirely rectify the fault. The root of the problem can be identified by examining column F; the feedback succeeds temporarily in correcting the output to the correct value, but the control loop contains no mechanism whereby the future of the input can be estimated and therefore the control loop reverts to the inactive state once the desired output has been achieved, and the system oscillates. A means of forecasting is required. The simplest forecast that can be made is that the input at $t + 1$ will be equal to the input at t, and this can be incorporated by modifying the equations to become.

7) $x_t = I_t - 2F_t$
8) $Q_t = 0.5\, x_t$
9) $F_t = F_{t-1} + (Q_{t-1} - G_{t-1})$

The starting trajectory then becomes as in Table 5.8.

TABLE 5.8

t	I	F	x	Q	G
0	10	-5	20	10	5
1	10	0	10	5	5
2	10	0	10	5	5
3	10	0	10	5	5
4	10	0	10	5	5
5	10	0	10	5	5

(See also Graph 5.16)

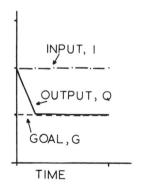

GRAPH 5.16 Starting trajectory for a modified system of Figure 5.3 (i.e. with prediction of future input).

This shows a marked improvement; output stabilises rapidly to the required value, and initial values are not critical. The trajectory for a fluctuation in input becomes as in Table 5.9.

TABLE 5.9

t	I	F	x	Q	G
6	10	0	10	5	5
7	11	0	11	5.5	5
8	12	.5	11	5.5	5
9	13	1.0	11	5.5	5
10	14	1.5	11	5.5	5
11	15	2.0	11	5.5	5
12	14	2.5	9	4.5	5
13	13	2.0	9	4.5	5
14	12	1.5	9	4.5	5
15	11	1.0	9	4.5	5
16	10	.5	9	4.5	5
17	10	0	10	5.0	5

(See also Graph 5.17.)

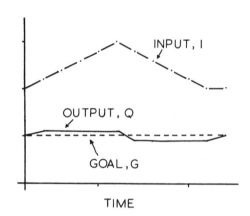

GRAPH 5.17.   Trajectory for a ramp input to the modified system of Figure 5.3.

This again shows a marked improvement. The control index becomes

$$C = (1 - (5.5 - 4.5)/2.5) \times 100\%$$
$$= 60\%$$

a three-fold improvement, and there is no tendency to oscillation. The trajectory for a step change in input becomes as in Table 5.10.

TABLE 5.10

t	I	F	x	Q	G
18	10	0	10	5	5
19	10	0	10	5	5
20	15	0	15	7.5	5
21	15	2.5	10	5	5
22	15	2.5	10	5	5
23	15	2.5	10	5	5

(See also Graph 5.18.)

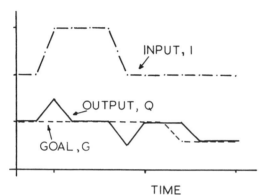

TIME

GRAPH 5.18   Trajectories for a step input and for a change in goal for the modified system of Figure 5.3.

This again shows a marked improvement. The response is extremely rapid, the output adjusts to the goal, and there is no tendency to oscillation. The same is true for a change in goal, as shown in Table 5.11.

TABLE 5.11

t	I	F	x	Q	G
24	10	2.5	5	2.5	5
25	10	0	10	5	5
26	10	0	10	5	5
27	10	0	10	5	3
28	10	2	6	3	3
29	10	2	6	3	3
30	10	2	6	3	3

(See also Graph 5.18.)

Here the same characteristics of rapid response, correct output, and no oscillation can be seen.

A question of some interest is the relative importance of an accurate model and of input prediction. An indication of this can be obtained by examining the trajectories of the system being investigated here when the predictive factor is retained but the model is less accurate. The equations then become

10) $x_t = I_t - F_t$
11) $Q_t = 0.5\, x_t$
12) $F_t = F_{t-1} + (Q_{t-1} - G_{t-1})$

The starting trajectory is as in Table 5.12.

TABLE 5.12

t	I	F	x	Q	G
0	10	−5	15	7.5	5
1	10	−2.5	12.5	6.25	5
2	10	−1.25	11.25	5.62	5
3	10	−.62	10.62	5.31	5
4	10	−.31	10.31	5.16	5
5	10	−.15	10.15	5.07	5
6	10	−.07	10.07	5.03	5
7	10	−.03	10.03	5.02	5
8	10	−.02	10.02	5.01	5
9	10	−.01	10.01	5.00	5
10	10	−0.00	10.00	5.00	5

(See also Graph 5.19.)

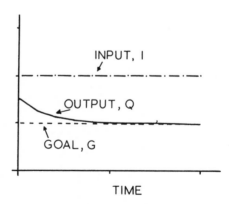

TIME

GRAPH 5.19.   Starting trajectory for the system of Figure 5.2 with input prediction.

The trajectory is now not critically dependent upon initial conditions, and there is no tendency to oscillation. The time to reach the goal is, however, considerably extended.

The trajectory for a fluctuating input is as shown in Table 5.13.

TABLE 5.13

t	I	F	x	Q	G
11	11	0	11	5.5	5
12	12	.5	11.5	5.75	5
13	13	1.25	11.75	5.87	5
14	14	2.12	11.88	5.94	5
15	15	3.06	11.94	5.97	5
16	14	4.03	9.97	4.98	5
17	13	4.01	8.99	4.49	5
18	12	3.50	8.50	4.25	5
19	11	2.75	8.25	4.12	5
20	10	1.87	8.13	4.07	5
21	10	.94	8.06	4.03	5
22	10	−.03	10.03	5.01	5
23	10	−.02	10.02	5.01	5
24	10	−.01	10.01	5.00	5
25	10	−0.00	10.00	5.00	5

(See also Graph 5.20.)

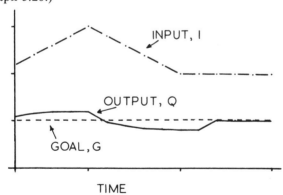

GRAPH 5.20.   Trajectory for a ramp input to the system of Figure 5.2 with input prediction.

The control index is

$$C = (1 - (5.87 - 4.03)/2.50) \times 100\%$$
$$= 27.4\%$$

a figure slightly greater than the index for a perfect model but no prediction. Again, the response is slow, and there is some tendency to oscillation, though this is well damped. The response to a step change in input is in Table 5.14.

TABLE 5.14

t	I	F	x	Q	G
26	10	0	10	5	5
27	15	0	15	7.5	5
28	15	2.5	12.5	6.25	5
29	15	3.75	11.25	5.62	5
30	15	4.37	10.63	5.31	5
31	15	4.68	10.32	5.16	5
32	15	4.84	10.16	5.08	5
33	15	4.92	10.08	5.04	5
34	15	4.96	10.04	5.02	5
35	15	4.98	10.02	5.01	5
36	15	4.99	10.01	5.00	5
36	15	5.00	10.00	5.00	5

(See also Graph 5.21.)

The response is slow, but accurate to the required value, and there is no tendency to oscillation. The picture for a step change in goal is as in Table 5.15.

TABLE 5.15

t	I	F	x	Q	G
37	10	0	10	5	5
38	10	0	10	5	3
39	10	2	8	4	3
40	10	3	7	3.5	3
41	10	3.5	6.5	3.25	3
42	10	3.75	6.25	3.12	3
43	10	3.87	6.13	3.07	3
44	10	3.94	6.06	3.03	3
45	10	3.97	6.03	3.01	3
46	10	3.98	6.02	3.01	3
47	10	3.99	6.01	3.00	3
48	10	4.00	6.00	3.00	3

(See also Graph 5.22.)

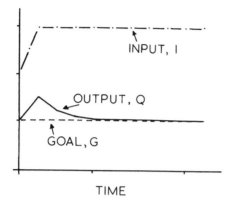

GRAPH 5.21.   Trajectory for a step input to the system of Figure 5.2 with input prediction.

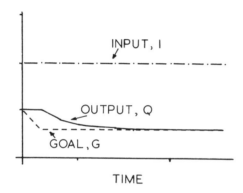

GRAPH 5.22   Trajectory for a change in goal in the system of Figure 5.2 with input prediction.

Again, the same picture emerges, a slow but accurate response with no oscillation.

The overall conclusion that emerges from this examination of an absolutely minimal feedback loop is that rapid response to an input depends upon having an accurate model, but that stability and accuracy of response depend upon prediction of future input within the feedback loop. Only the most elementary form of this prediction has been considered, but its value has been clearly demonstrated.

This value was in part due to the extremely simple forms of input considered. More complex inputs would require more complex prediction functions, involving rates-of-change calculations. Such factors are commonly included in the design of servo control systems, for reasons of stability (see, for example, Distefano (1967) Chapter I, Chapter 5) but the reason for their inclusion is given as to provide damping in the system. Although the basic mathematics remain the same, the reason that emerges from the foregoing line of reasoning is different; it is to allow for the prediction of the input.

In the context of managerial systems, the above line of reasoning leads to some interesting conclusions. It suggests that there are three distinct elements to administrative management, namely the obtaining of information about output, the modelling of the system under control, and the prediction of future input. Furthermore, these latter two have quite distinct areas of importance. An accurate model of the system is important for speedy response, but in isolation it produces instability and inaccuracy in the controlled output. Prediction of the input allows stability and accuracy of response.

Naturally, optimum results are obtained with a combination of the two but the possibilities for trade-off between the two are limited, since they affect different factors. Furthermore, it would seem that the ability to predict an input is of relatively greater importance than the possession of an accurate system model, in that a smooth and accurate response is more dependent upon this than upon detailed knowledge of the system. As far as it goes, this offers support for the view that a good manager can manage any operation with a high level of success; he needs only a very approximate knowledge of the operation under his command, provided that speed of response is not vital.

From a theoretical standpoint, it is of interest that nominally feedback systems can (and by implication usually do) contain predictive elements. It adds weight to the view expressed earlier that the main purpose of strategic control is not to obtain faster response through a prediction of future input, but is much more concerned with problems of overall organisation.

Thus far, only the most elementary of feedback situations has been examined. As has been emphasised, a managerial situation involves multiple goals and simultaneous control of several variables. It is relevant, therefore, to examine more complex situations, particularly where variables are not separable but interact. As an archetype of such a situation, consider Figure 5.4 with equations.

13) $x_t = A_t - 2F1_{t-1}$
14) $Q_t = \cdot 5x_t + \cdot 5y_t$
15) $F1_t = F1_{t-1} + Q_{t-1} - G1_{t-1}$
16) $y_t = B_t - 2F2_{t-1}$
17) $R_t = \cdot 5y_t + \cdot 5x_t$
18) $F2_t = F2_{t-1} + R_{t-1} - G2_{t-1}$

Here there is strong interaction of the variables, but each control loop assumes that the variables are independent. Each control loop has a predictive

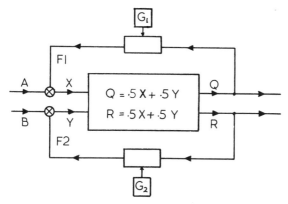

FIGURE 5.4. A more complex model of feedback activity, with interacting variables and multiple feedback.

element, and an accurate model of the effect of the variable it controls. These features make the situation described somewhat unrealistic, but serve to emphasise the principles involved. The trajectory for a varying input is as shown in Table 5.16.

TABLE 5.16

t	A	B	F1	F2	x	y	Q	R	G1	G2
0	10	10	0	0	10	10	10	10	10	10
1	11	10	0	0	11	10	10.5	10.5	10	10
2	12	10	.5	.5	11	9	10.0	10.0	10	10
3	13	10	.5	.5	12	9	10.5	10.5	10	10
4	14	10	1.0	1.0	12	8	10.0	10.0	10	10
5	15	10	1.0	1.0	13	8	10.5	10.5	10	10
6	14	10	1.5	1.5	11	7	9.0	9.0	10	10
7	13	10	.5	.5	12	9	10.5	10.5	10	10
8	12	10	1.0	1.0	10	8	9.0	9.0	10	10
9	11	10	0	0	11	10	10.5	10.5	10	10
10	10	10	.5	.5	9	9	9.0	9.0	10	10
11	10	10	-.5	-.5	11	11	11.0	11.0	10	10
12	10	10	.5	.5	9	9	9.0	9.0	10	10
13	10	10	-.5	-.5	11	11	11.0	11.0	10	10

(See also Graph 5.23.)

The control index (for a single variable) is

$$C = (1 - (11.0 - 9.0)/2.5) \times 100\%$$
$$= 20\%$$

which is not very high. Response is rapid, as would be expected, but the output enters a cycle. The trajectory for a change in one goal is shown in Table 5.17.

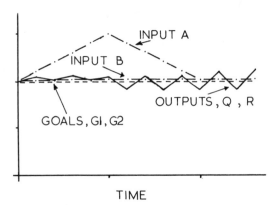

TIME

GRAPH 5.23   Trajectories for a single ramp input to the system of Figure 5.4.

TABLE 5.17

t	A	B	F1	F2	x	y	Q	R	G1	G2
0	10	10	0	0	10	10	10	10	10	10
1	10	10	0	0	10	10	10	10	6	10
2	10	10	4	0	2	10	6	6	6	10
3	10	10	4	-4	2	18	10	10	6	10
4	10	10	8	-4	-6	18	6	6	6	10
5	10	10	8	-8	-6	26	10	10	6	10
6	10	10	12	-8	-14	26	6	6	6	10
7	10	10	12	-12	-14	34	10	10	6	10
8	10	10	16	-12	-22	34	6	6	6	10
9	10	10	16	-16	-22	42	10	10	6	10

(See also Graph 5.24)

The system never settles, but oscillates between the two goals. A feature of interest is the ever-increasing feedback activity involved. The trajectory for a step change in input is as shown in Table 5.18.

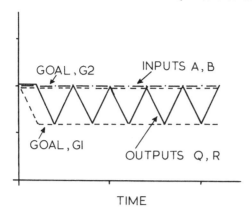

GRAPH 5.24   Trajectories for a change of one goal in the system of Figure 5.4.

TABLE 5.18

t	A	B	F1	F2	x	y	Q	R	G1	G2
0	10	10	0	0	10	10	10	10	10	10
1	15	10	0	0	15	10	12.5	12.5	10	10
2	15	10	2.5	2.5	10	5	7.5	7.5	10	10
3	15	10	0	0	15	10	12.5	12.5	10	10
4	15	10	2.5	2.5	10	5	7.5	7.5	10	10
5	15	10	0	0	15	10	12.5	12.5	10	10
6	15	10	2.5	2.5	10	5	7.5	10	10	10

(See also Graph 5.25.)

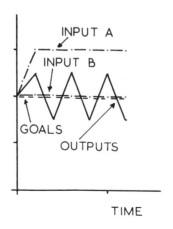

GRAPH 5.25.   Trajectories for a step change of one input in the system of Figure 5.4.

Again, the response enters a cycle, oscillating around the goal values but in this instance the amount of feedback activity does not increase.

It is perhaps worth pointing out here that it is not meant to infer that the system is unstable because its output cycles; the cycle itself is quite stable. However, from the point of view of a manager, such cycles may well suggest that his operation is out of control, especially as in real life the cycles will not be so clear-cut as in these grossly simplified examples. Furthermore, other members of the organisation, who receive the output of the managers department as input to their own operations, will not be satisfied with such cycles.

Since the feedback loops described already have input prediction methods that have been found adequate, it is logical to look at the system models in the loops to attempt to improve the situation. The models used ignored the interaction between variables. To improve them, this needs to be taken account of — but only in one of the loops. If the equations are modified to become

19) $x_t = A_t - 2F1_t$
20) $Q_t = \cdot 5x_t + \cdot 5y_t$
21) $F1_t = F1_{t-1} + (Q_{t-1} - G1_{t-1}) - (R_{t-1} - G2_{t-1})$
22) $y_t = B_t - 2F2_t$
23) $R_t = \cdot 5x_t + \cdot 5y_t$
24) $F2_t = F2_{t-1} + (R_{t-1} - G2_{t-1})$

which models the interaction in the $G1$ control loop, the trajectory for a disturbed input becomes as shown in Table 5.19.

TABLE 5.19

t	A	B	$F1$	$F2$	x	Y	Q	R	$G1$	$G2$
0	10	10	0	0	10	10	10.0	10.0	10	10
1	11	10	0	0	11	10	10.5	10.5	10	10
2	12	10	0	.5	12	9	10.5	10.5	10	10
3	13	10	0	1.0	13	8	10.5	10.5	10	10
4	14	10	0	1.5	14	7	10.5	10.5	10	10
5	15	10	0	2.0	15	6	10.5	10.5	10	10
6	14	10	0	2.5	14	5	9.5	9.5	10	10
7	13	10	0	2.0	13	6	9.5	9.5	10	10
8	12	10	0	1.5	12	7	9.5	9.5	10	10
9	11	10	0	1.0	11	8	9.5	9.5	10	10
10	10	10	0	.5	10	9	9.5	9.5	10	10
11	10	10	0	0	10	10	10.0	10.0	10	10

(See also Graph 5.26.)

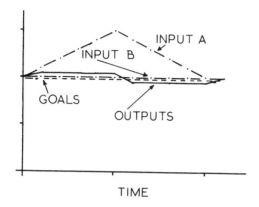

TIME

GRAPH 5.26   Trajectories for a ramp input to the modified system of Figure 5.4.

The control index is

$$C = (1 - (10.5 - 9.5)/2.5) \times 100\%$$
$$= 60\%$$

showing a good level of control. There is no tendency to oscillation. Examination of the $F1$ column (the feedback with the accurate model) shows that there is no activity from this loop — it has been 'shorted out' so to speak, with considerable benefits. The trajectory for a step change in input is similarly improved, as shown in Table 5.20.

TABLE 5.20

t	A	B	$F1$	$F2$	x	y	Q	R	$G1$	$G2$
0	10	10	0	0	10	10	10	10	10	10
1	15	10	0	0	15	10	12.5	12.5	10	10
2	15	10	0	2.5	15	5	10	10	10	10
3	15	10	0	2.5	15	5	10	10	10	10
4	15	0	0	2.5	15	5	10	10	10	10
5	15	10	0	2.5	15	5	10	10	10	10

(See also Graph 5.27.)

Again, a well-controlled response, with $F1$ showing every sign of masterly inactivity. The response to a change in goal is as in Table 5.21.

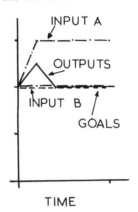

GRAPH 5.27   Trajectories for a step input to the modified system of Figure 5.4.

TABLE 5.21

t	A	B	F1	F2	x	y	Q	R	G1	G2
0	10	10	0	0	10	10	10	10	10	10
1	10	10	0	0	10	10	10	10	6	10
2	10	10	4	0	2	10	6	6	6	10
3	10	10	4	0	2	10	6	6	6	10
4	10	10	12	-8	-14	26	6	6	6	10
5	10	10	16	-12	-22	34	6	6	6	10

(See also Graph 5.28.)

The situation for a change in $G2$ is slightly different, as shown in Table 5.22.

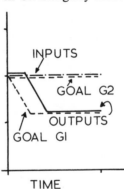

GRAPH 5.28.   Trajectories for a change in G1 in the modified system of Figure 5.4.

In the first case, $G1$ is achieved (at the expense of $G2$) but is only maintained by rapidly increasing feedback. If $F1$ and $F2$ are limited, as is normally the case, the situation will reach oscillation. In the second case, a cycle sets in, and neither target is reached, although both feedback channels are active.

TABLE 5.22

t	A	B	F1	F2	x	y	Q	R	G1	G2
0	10	10	0	0	10	10	10	10	10	10
1	10	10	0	0	10	10	10	10	10	6
2	10	10	4	4	2	2	2	2	10	6
3	10	10	0	0	10	10	10	10	10	6
4	10	10	4	4	2	2	2	2	10	6
5	10	10	0	0	10	10	10	10	10	6

(See also Graph 5.29.)

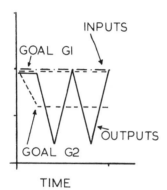

GRAPH 5.29.   Trajectories for a change in G2 in the modified system of Figure 5.4.

The conclusions to be drawn from the above examples seem to be as follows. In the first place, where a system contains interacting variables, an accurate model of it is required for acceptable control. In the second place, attempting joint control of all variables does not necessarily lead to improvements in overall control, and may well impair results rather than improve them. This is of relevance to the design of organisational control systems. In the third place, attempting to change the goals of such a system without at the same time modifying the system is fraught with difficulty. In other words, a multi-variable system will not behave as a servo-mechanism.

Table 5.21 contains a point of considerable relevance here, to which reference has already been made. Goals may be maintained for a period of time by the expenditure of considerable feedback effort, until the available resources are exhausted. There will then be a sudden step change in output, with little or no apparent immediate cause. Industrial relations problems have these characteristics, and it may be that some such mechanisms are involved in these circumstances. The earlier discussions on organisational style and conflict between personal and organisational goals are also of relevance here.

This leads to a further point of relevance. Reference has been made to the importance of accurate modelling for control purposes. Yet, by definition, organisations contain people, and accurate models of human behaviour are not available. Thus, the strictly deterministic equations that have been used in the above examples are not truly representative of the managerial situation. This does not, however, invalidate the nature of the conclusions drawn, which are qualitative rather than quantitative in nature.

Thus far, only extremely simple cases have been examined. Enough has been said, however, to establish the complexity of interrelationships that exists within organisations, and the problems of control and stability that this entails. It is also clear that Ashby's (1960) concept of a multi-stable system is applicable to organisations. (Indeed, this could have been assumed from the start, but it appeared advisable to establish the identity rather than to assume it.) Ashby has studied the general problem of stability in such systems ((1960), Chapter 20) and concluded (though not proved) that the probability of stability decreases as the number of variables increases. He says (p.261), "These results prove little; but they suggest that the probability of stability is small in large linear systems assembled at random." Porter (1972) has come to similar conclusions. This does not mean that large systems are necessarily unstable — the prolonged existence of many organisations demonstrates this — but it does imply that such stabilities are not easily found, and are easily disturbed. This may well be the instinctive reasoning behind the conservatism of many large organisations.

Ashby's work also suggests a reason for the departmentalisation found in many large organisations. His work on adaptation (1960, Chapter 16) shows that the time taken for a multistable system to adapt to its environment can be decreased by many orders of magnitude if the total is partitioned into subsystems, with minimal communication between them. Many organisations are in fact patterned in this way (e.g. Buying Division, Production Division, Sales Division, with further subdivisions in each), though it does not, of course, follow that such partitions necessarily correspond to operational reality. Further, it suggests strongly that attempts to break down such organisational barriers and encourage communication are ill-advised and may be strongly counter-productive.

It also adds evidence to the previous conclusion that there is a clear distinction between administrative and strategic management. The former has to run the organisation as it exists; the latter has to reshape and re-structure it as the environment changes. This distinction is often not made clearly in actual organisations, where the same group of individuals carry out both functions. The required distinction is akin to the distinction between line and staff management, but with clearer responsibilities and more authority accorded to the staff.

5.3.2 Consistency of models among managers

Considerable discussion has been afforded already to the importance of models in the managerial process. One aspect of this topic which has not been examined is that of consistency amongst the different models used by different managers in the same organisation, and it merits a brief examination here.

It is well-known that the same phenomenon (or system) can be modelled in a variety of incompatible ways, yet each model yields valid results. For example, light can be modelled as a wave process, or as a particle process — these two models have only recently been reconciled one with another. McGregor's (1960) Theory X and Theory Y form another such pair. Thus, it is to be expected that two managers confronted with the same situation may model it in entirely different ways, each of which can be valid.

To some extent — particularly in the administrative situation, where only a black box model is required — this is of no great consequence. However, it is well-known that organisations tend to develop their own "style". The most obvious examples of this are perhaps the "City Gent", the "Civil Service Mind" or the "Military Manner". This implies a certain degree of consistency amongst the managerial models in use in a particular organisation, and it is of interest as to how this uniformity develops, and what some of the consequences may be.

The most obvious solution to the problem of how uniformity develops lies in a process of "natural selection", akin to Ashby's "selection by equilibrium". For managers interact with each other, both formally and informally, and communication will be easier amongst those with similar models and hypotheses. Groups will tend to coalesce and cohere on the basis of similarity of models.

A case of particular importance is that of promotion. Other things being equal, the individual who thinks like his superiors is more likely to be appointed to a senior position, on the basis that he will fit in better with colleagues, will more easily form part of a team, and will be less likely to "rock the boat". Indeed, it could even be that such considerations might outweigh considerations of merit and achievement – particularly when it is extremely difficult to measure managerial performance in any meaningful way.

Thus there is a natural tendency for organisations to select individuals who conform to the organisational pattern. People with widely dissenting views then find themselves as misfits, and either leave or make no further progress. Ultimately, the recruiting procedures are likely to reject such individuals before they even join the organisation. Conversely, there will be the opportunity for people who fit in exceptionally well to make rapid progress, almost regardless of ability. Since "fitting in" is in part a function of background and education, it is to be expected that large organisations will develop well-defined "styles" of their own.

There is a danger in this process. As has been mentioned, a wide variety of models can yield successful results over a limited range of circumstances. If the actual situation should move outside the range where the "entrenched model" operates satisfactorily, then the organisation will find it more difficult to adapt to the new conditions. This is partly because there is no internal source of a radically different view which can provide the seed for change, and partly because the communication between members of management will tend to reinforce the view that what is happening is only a temporary aberration and things will soon be back to normal. If the organisation in question is sufficiently large and powerful, it may even attempt to adjust the outside world back to the situation that used to prevail.

It can prove difficult for an organisation to get out of this situation once it has arisen. In the commercial sphere, the problem will probably be resolved by the firm going out of business; with some other types of organisation, the state of affairs may last almost indefinitely.

5.3.3 The modelling process

Much attention has already been directed to models, but little has been said about the ways in which they may be developed, apart from the fact that information outside normal feedback channels is used. This reflects the paucity of attention that has been paid to the subject within management literature. Whilst cybernetic literature deals quite specifically in models, again there is comparatively little attention to the process of how models are arrived at. For example, though Ashby (1956) examines the way that scientific models are developed – and pays great attention to how state-determined system models are developed – he does look at the basic problem of how the initial variables are selected for study. This is a crucial problem, which Ashby is apparently prepared to leave to chance.

This initial selection of variables is a deep problem and will be discussed here. For the moment, Ashby's starting point will be taken, and related to the more immediate practical context of managerial modelling.

First, let it be admitted that there are specialists in modelling in management sciences such as OR. Much of this type of work lies in developing models and analysing them. These, however, are not the immediate concern – they are too cumbersome to be used in the day-to-day hurly-burly of management life. The concern is more with the basis for the immediate 'seat-of-the-pants' control.

First, there is a point to be made about the general nature of organisational work at the operational level. Tasks are designed to be performed in isolation from each other, i.e. the total system is generally assumed to be serial in nature. There is seldom any intrinsic feedback between the tasks themselves, there is not the sort of complex interrelation found in the organic, or biological or ecological fields. The tasks are designed to be regulated by external feedback. As has been seen, the complexity of organisations arises from interrelations amongst these external feedback mechanisms rather than direct interaction amongst the operations themselves.

Such design facilitates enormously the problem of controlling the organisation. In passing, it should be noted that the sphere of government does not accord with this principle; a nation is much more analagous to an organic entity that to a serial process, and the basic task of government is not so much to ensure that x amount of goods and services are produced as to ensure that there is a proper set of checks and balances available to preserve the viability of the social order. Because the social order is itself a complex interaction between individuals and groups of individuals, it would appear that the basic philosophy of government should be different from the basic philosophy of organisations.

However, the manager in an organisational context is faced basically with an input-output situation which he is required to control. For this, he requires an input-output type of model. If he were starting from scratch, he could of course use the methods described by Ashby. Usually of course, he is not in this position. He comes equipped with a variety of models, one of which he will select as appropriate and proceed to use.

These models will almost always contain the equivalent of adjustable parameters (such as feedback fraction, delay time, inertia) the values of which will need to be adjusted to the particular situation. Thus in a parallel-management operation (such as Sales Management) it may well be that the sales manager has one basic model of his retail outlets which is adapted to each one by the substitution of appropriate values for parameters such as outlet size, number of different type of goods sold, number of staff, and so on. (This is not necessarily the case; it may be that a single set of parameters is used for all. There is no guarantee that a manager will do what theory prescribes.)

The ways in which such 'tuning' may be achieved is well understood, and have been described by Garner (1968) for electronic systems. The technique is to apply the same input to both system and model, and adjust the model parameters until identical outputs are achieved, i.e. basically the model and system are run in parallel. In the present context, however, this implies that the model cannot be used for control purposes whilst this "tuning" is carried out, or the interaction would produce undecideable results. The manager cannot afford to spend too long "tuning" before he starts on his job.

It can of course be the case that the form of a model and the form of the reality are quite different, and yet by suitable adjustment of the parameters the input-output relationship may be identical, or at least approximately equal over the range of inputs studied. As a simple example the functions

$$y = \sin x$$

can be modelled by

$$y = x$$

for small values of x. Thus a manager may select an entirely inappropriate model, yet operate successfully — at least over a limited range of input.

Should the input vary beyond this range, the manager may find himself with serious problems, and locked in to a very difficult situation. In his attempts to exert control through an inappropriate model, he will find himself using higher and higher levels of feedback activity. It is easy to say that what he needs to do is to stand back from the situation, select a new model, and adjust its parameters to the situation. Yet, the situation is probably such that to temporarily abandon control (which is what is implied) could have major consequences. Furthermore, his experience has taught him that his model is right — it has always worked before — and thus he is psychologically reluctant to abandon it. He may well feel that a step change has occurred in the environment, though he is unlikely to use that terminology to express his views.

The probable sources of a manager's initial models are fairly obvious, either training or experience. Training will have equipped him with a variety of theoretical models of the process of which he is to be in charge. Alternatively, experience may have taken him through positions as an operator of various sections of the process. This latter will have enabled him not only to "tune" his model to the process but also to model the influence of the people in the process, an element not present in the theoretical background. A variant of this situation is where a manager is recruited from outside the organisation; a quite usual requirement for appointing a recruitee is experience in a similar position. This is particularly true for senior positions in management. Whilst the reasons for this requirement are obvious, it has its dangers. A nominally similar job in a different organisation, being in a totally different situation, may have produced models that are not applicable in the new organisation. The process is in some ways analagous to tissue transplants; the body may well reject the graft.

Brief mention was made above of modelling the people involved in the process. There are two aspects to this. The first is to model the performance of the worker whilst performing his duties – this can be done by observing the performance of the total man/task system. This, however, will provide only extremely limited information about how he will perform on other duties. To do this, a model of the person himself is required, which can only be acquired through social interaction. This point is worthy of comment chiefly because of the efforts which many organisations make to limit any such interaction between different levels in the organisational hierarchy. The whole apparatus of status and position is brought into play, in an attempt to ensure that people interact only in their organisation's roles and not as people. Whilst this acts to preserve the organisation as a serial process rather than an organic whole, and thus ensures ease of control, it has serious implications for the extent to which individuals can expect to achieve their personal goals – or indeed to maintain their self-respect. It also has implications for the speed at which change can be implemented, i.e. how rapidly the organisation can respond to the environment. Large bureaucratic organisations need to plan in more detail, being less able to depend upon individual initiative to cope with new situations. On the other hand, organisations only become large and bureaucratic in a basically stable environment.

It has been stated earlier that, for administrative management, only a "black box" (i.e. input-output transform) model is required, though it was not asserted that these are necessarily what is used. However, for strategic management, it is necessary to have a more detailed model showing how the various components interact. This is because strategic management involves re-shaping the organisation, either rearranging its constituent parts, replacing them with new ones, or adding further operations. This cannot be done adequately without some knowledge of how the existing organisation is put together and what the potentials of these parts are.

Developing this type of model is rather different from developing a black box model; essentially, it consists of stringing a series of black boxes together, the characteristics of each of which are known. (The possession of this type of model is, of course, equivalent to being able to explain the process under consideration.)

Generally, such models cannot be inferred simply from knowledge of the overall transfer function. (Because, to repeat the quotation from Ashby and Conant (1970), "almost anything may serve as a model for almost anything else" in the context of black box models.) It must be built up from knowledge (or special study) of the internal functioning of the organisation.

This is because the final model must bear structural similarity to the real situation. This in turn, is because the planning process is, in essence, to permute the structure of the model, changing the connectivity between sub-functions, changing the nature of some of the sub-functions, or adding (or deleting) sub-functions. The overall transform of each permutation is then predicted, and the optimum one for a given set of goals is chosen. (Or perhaps, a satisfactory rather than optimum solution may be sought.)

Obviously, the number of permutations available depends upon the number of identified sub-functions or, in other words, the degree of detail in the model. Frequently, there are heuristics available to limit the number of permutations examined by indicating which set are unlikely to yield reasonable results. These are necessary to limit the search time to reasonable bounds. Such heuristics are not infallible, and significant advances can sometimes be made by abandoning them and searching through combinations not previously examined. This may be described as "lateral thinking" or as the breaking of preconceived ideas. Beer (1966) gives a good example of this in his description of the project to re-site a production location, with the attendant effects upon a distribution network.

Also of interest in this example is the amount of work it entailed, both in the initial preparation of the model and the subsequent computation performed upon it. (It is also of interest that Beer does not discuss how the model was validated.) It illustrates very well the amount of computation required for full exploration of even a relatively simple model, at the level of factories and depots. Usually, such time is not available. This consideration is, however, leading on to a topic more fully explored in the subsequent section.

Before leaving the topic of the modelling process, it should be mentioned that in practice it is made more difficult by what may be collectively described as 'noise'. This can take several forms. The most obvious is noise in the classical sense of errors in figures, and reports. Another is what may be termed 'false impressions'. These may arise either from attempts by subordinates to show themselves in a good light, or by what may be thought of as a sampling error, in that the manager observes various parts of his operations intermittently rather than continuously. A third form is imposed by the managers perceptual limitations; he will filter the information he receives and may, in so doing, distort it.

All these factors make the process of modelling more difficult, in addition to which the variable nature of the tasks being performed must be considered. The performance of an organisational system is essentially statistical in nature, a compound of peaks and troughs in demand, variability in raw materials and operations, breakdowns, and other unforeseen occurrences.

In principle, such factors present no great problem (once they are recognised) apart from time. Given a sufficiently long sample of behaviours they can be allowed for. However, in organisational life, such time is not always available. The result will be less accurate models at managerial level.

5.3.4 Speed of data processing

The question of speed of data processing has occurred previously, in the context of channel capacity. There it was argued that the required channel capacity was determined by the speed at which the models used could process information. This is an extremely important topic, which is examined in greater depth in this section.

It is as well to start with a reminder of the purpose of processing information through a model. It is to gain control, either of the environment or of an organisation. An important point to remember is that this control does not have to be absolute, in the sense that it may not be essential to maintain the controlled variable at a precise level; it may well be sufficient to ensure that it does not exceed prescribed and fairly wide limits, or even that it does not fall below a certain critical value. This is perhaps most clearly seen in relation to controlling the environment. Obviously the organisation cannot expect to exert close control over all the features of the environment, yet it may be able to gain great advantage from being able to bring to bear a limited amount of influence over some of them. A case in point, for business organisations, is furnished by the various forms of sales tax; whilst firms cannot expect to exercise absolute control over such taxes, it is clearly in their interests to keep them as low as possible.

It follows, then, that what is important is not so much that the model yields precise and detailed results as that it processes information rapidly enough to enable a response to be made in time for it to be effective. A timely response in the right general direction is preferable to a more accurate response too late to be effective.

Therefore, it is relevant to look at the factors that govern the computational speed of a model. Consider first a simple two-input one-output model as shown in Figure 5.5. This is the simplest form of model that can be constructed; in its very simplest form, one of the inputs is held constant.

The first stage in computing the output, y, is to evaluate the inputs A and B. This is equivalent to placing them in categories. Suppose input A can be classified into x_1 categories, input B into x_2 categories. Assuming that x_1 and x_2 are ordered sets, and that a "split half " technique can be used, this will require

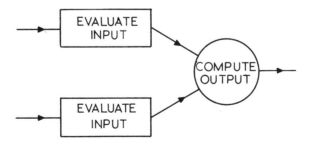

FIGURE 5.5.   The structure of a simple model.

$$\log_2 x_1 + \log_2 x_2$$

computations. Evaluation of the output is equivalent to looking up a cell in an $x_1 \times x_2$ table†, which will require a further $\log_2 (x_1 \times x_2)$ computations. Thus the total number of computations required, N, is given by

$$N = \log_2 x_1 + \log_2 x_2 + \log_2 (x_1 \times x_2)$$
$$= 2 \log_2 (x_1 \times x_2)$$

For P inputs, N is given by

$$N = 2 \log_2(x_1 \times x_2 \dots x_n \dots x_p)$$
$$= 2 \sum_1^p \log_2 x_n$$

If the model is in two stages, α and β as shown in Figure 5.6, then the stages will require

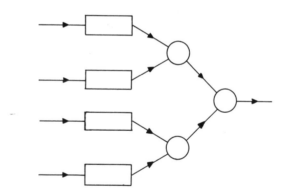

FIGURE 5.6.   The structure of a more complex model.

$$N_1 = 2 \sum_1^{p_\alpha} \log_2 x_n + 2 \sum_{p_\alpha+1}^{p_\beta} \log_2 x_n$$

$$= 2 \sum_1^p \log_2 x_n$$

---

† This assumes there is no computable relation between input and output. The case where there is a computable relation is considered later.

computations. The final stage will require

$$N_2 = \log_2(y_1 \times y_2 \ldots y_n \times y_q)$$

calculation, where there are q inputs to the final stage.

Since $\quad y_1 = x_1 \times x_2 \ldots x_a$
$\qquad\quad y_2 = x_{a+1} \times \ldots x_b$

etc,

$$N = \log_2(x_1 \times x_2 \ldots x_n \ldots \times x_p)$$

Hence $\quad N_2 = N_1 + N_2$

$$= 3 \times \sum_1^p \log_2 x_n$$

and for an f-stage model,

$$N = (f+1) \sum_1^p \log_2 x_n$$

If the mechanism that is running the model is capable of K computations per second, then the rate R at which output signals can be produced is

$$R = K/N$$
$$= \frac{K}{(f+1)} \cdot \frac{1}{\sum_1^p \log_2 x_n}$$

In the above, it has been assumed that there is no computable relationship between input and output, and y has to be looked up in a contingency table. If there is a computable relationship, i.e. if

$$y = f(x_1 ; x_n)$$

that requires only F computations, then the value of N will obviously change. In the first example, we have

$$N = F + \sum_1^p \log_2 x_n$$

and for the more general case, assuming that each mode requires Fa computations

$$N = \sum_{1}^{\beta} Fa + \sum_{1}^{p} \log_2 x_n$$

for β nodes. The response rate becomes

$$R = K/N$$

$$= K/(\sum_{1}^{\beta} F_\alpha + \sum_{1}^{p} \log_2 x_n)$$

In order to maximise R, with a fixed value of K (i.e. with a given computing mechanism), N must be minimised. Here there are two cases to consider, firstly where there are no computable functions, secondly where there are the
In the first case we have

$$N = (f+1) \sum_{1}^{p} \log_2 x_n$$

and this implies three possible strategies:

1) minimise x_n, i.e. use fewer (and therefore possibly broader) categories to classify the orginal inputs.

2) minimise p i.e. restrict the number of inputs used to the model.

3) minimise f, i.e. reduce the number of stages in the model.

It is of relevance here to examine the relative efficiencies of each of these strategies. To illustrate this, assume an initial model with 10 stages, 10 inputs, each of which is classified into 8 categories.
This will require

$$N = 11 \times 10 \times 3$$
$$= 330 \text{ computations.}$$

If the number of stages is halved, this becomes
$$N' = 6 \times 10 \times 3$$
$$= 180 \text{ computations.}$$

If the number of inputs is halved, this becomes

$$N'' = 11 \times 5 \times 3$$
$$= 165 \text{ computations.}$$

If the number of categories is halved, this becomes

$N''' = 11 \times 10 \times 2$
$= 220$ computations.

Thus in general, (except for the special case where $n < 4$), the maximum effect on R is gained by reducing the number of inputs to the model. An almost equal effect is gained by reducing the number of stages. The least effect is gained by reducing the number of input categorisations. (These examples ignore any interactions between f, p, and n. Thus, in particular, reducing the value of p may enable simplification of the model to take place, with consequently greater effect. Such interactive effects will depend upon the specific model being used.)

The situation where the model consists of computable functions is different. The problem is to minimise.

$$N = \sum_{1}^{\beta} F a + \sum_{1}^{p} \log_2 x_n$$

Here there are four basic strategies,

1) minimise β – i.e. reduce the number of stages in the model.

2) minimise α – i.e. reduce the number of computational steps required for each calculation.

3) minimise P – i.e. reduce the number of inputs.

4) minimise n – i.e. reduce the number of input categories. (It should be noted that n here is at least partly determined by the requirement of F_α. To operate in a decimal system, n will need to be at least ten.)

For purposes of illustration, assume again a model with 10 inputs, 10 stages, and each input classified into 8 categories. (This transgresses the above requirement for n, but it enables comparisons to be drawn more easily with the previous illustration.) Assume further that each calculation is of the simple type

$Y = ax + bz$

requiring 3 computations. Then we have

$N = 10 \times 3 + 10 \times 3$
$= 60$ computations.

If the number of stages is halved, this becomes

$$N' = 5 \times 3 + 10 \times 3$$
$$= 45 \text{ computations.}$$

If the number of inputs is halved, this becomes

$$N'' = 10 \times 3 + 5 \times 3$$
$$= 45 \text{ computations.}$$

If the number of categories is halved, this becomes

$$N''' = 10 \times 3 + 10 \times 2$$
$$= 50 \text{ computations.}$$

The most immediately striking feature of this illustration is the greatly reduced number of computations required compared to the earlier example. The reason for this is that the second approach utilises redundancies in the environment that were not employed in the first approach. Thus it can be concluded that the search for such redundancies (i.e. laws of nature) is worthwhile.

It was pointed out that the use of computable functions might well require the use of finer categorisation of the input. It is of interest to ask how many categories can be used before the computational advantages are lost, i.e. for what value of n is

$$\sum_{1}^{\beta} Fa + \sum_{1}^{p} \log_2 x_n = (f+1) \sum_{1}^{p} \log_2 x_n$$

Using values from the above examples, we have

$$10 \times 3 + 10 \log_2 x_n = 11 \times 10 \times 3$$

Hence $\log_2 x_n = 27$,

and $n = 2^{27}$ categories

a number sufficiently large for most practical purposes.

It is also of interest to enquire how complex can individual calculations be before the advantages of the method are overcome. Again using the above values for illustration, we have,

$$10.6 + 10 \times 3 = 11. \ 10.3$$

where α is the average number of computations required per calculation. The above expression yields

$$\delta = 30$$

which is sufficient to cope with, for example, polynomials up to the quartic of the form

$$Z = ax^4 + bx^3 + cx^2 + dx + ey^4 + fy^3 + gy^2 + hy + i.$$

Apart from these considerations, the example serves to illustrate that, in order to increase R, the most effective strategy is either to decrease the number of stages or the number of inputs. Decreasing the number of input categories has less effect than either of these two.

In the organisational situation, a high value of R is of great value. The pressures on managerial time are considerable, and the ability to make a reasonable decision quickly is often of more value than the ability to make an optimal response slowly. Thus it can be concluded that there are pressures that will drive managerial models towards having few stages, a restricted range of input and only broad discriminations of the input – in other words, towards simple black box models.

It is thus to be expected that management thought about organisations will be of an apparent simplicity, showing a certain roughness and crudeness of approach. They will tend to be stereotyped responses, classifying the world into extremely simple chains of cause-and -effect relationships, such as "lower price leads to higher sales" or "high morale leads to high productivity". This is no reflection on the general level of intelligence and sophistication of managers, but rather a consequence of the fact that these models have evolved as working tools for a specific job.

However, the approximate, black box nature of such models has certain theoretical consequences. In the first place, it has been shown previously that good control, in the sense of a high control index, was dependent upon having an accurate model of the system being controlled. A simple model and an accurate model are not necessarily incompatible requirements, but there are obviously difficulties in reconciling the two. The quality of a manager (in the administrative aspect of his job) may well be a function of how well he can achieve this reconciliation.

In the second place, it has been argued that models for strategic control need the opposite characteristics. They need to be accurate and detailed. This contributes to the line of reasoning that suggests that administrative and strategic management are different in nature, and should be more clearly and definitely defined in organisations.

It can also be suggested that organisations should seek to take more advantage of the power available in computable models. Technologically, with the advances in computing power now available, this is feasible. The need is for more quantitative modelling. This may well be difficult, especially in the field of industrial relations, but the effort seems potentially worthwhile.

Much of the above argument has been based on the need to maximise the rate of output. A question of interest is how is the computation affected when the rate of input exceeds the rate at which the model can compute. Experimental evidence that this does occur, and the effect this has on control, has already been presented.

To examine this question, it is convenient to use the formulation of Porter (1972), following on from the earlier work of Ashby (1960). The control situation is characterised in matrix terms, with the general form

$$Y = AX_t + BZ_t$$

where Y is the output, X_t is the input, A is the transfer matrix of the uncontrolled system, Z_t is the control input and B the control transfer matrix. (Thus B is equivalent to a model of the uncontrolled system. In what follows, it is assumed that the model is analytic, for the purpose of illustration.) The problem of control can then be considered as the problem of the computation of BZ_t. What is in question is how this computation is affected by overload of the input.

To take a simple example; suppose BZ_t contains only three components

$$Z_1 = a_1 y_1 + b_1 y_2 + c_1 y_3$$
$$Z_2 = a_2 y_1 = b_2 y_2 \quad b_2 y_2 + c_3 y_3$$
$$Z_3 = a_3 y_1 + b_3 y_2 + c_3 y_3$$

(where the t subscript has been omitted for convenience). Before overload occurs, the supposition must be that the control mechanism has sufficient computing power to calculate this function in each time interval $t_{n+1} - t_n$. When overload occurs, it is no longer able to do this, and must resort to strategies that allow approximate solution to be obtained. The possibilities seem to be as follows:

1) Compute BZt at longer intervals. This will allow Y to vary over a wider range, and consequently more vigorous control will be required. As the interval between computations gets longer, the control mechanism will approximate more closely to an on-off device.

2) Round off values of other constants or variables, to produce approximate rather than exact values. It should be noted that the value of Zt can be very sensitive to the values of the constant terms, and therefore such approximations may produce values well wide of the correct solution.

3) Combine elements together and treat, e.g. Z_1 and Z_2 as a single variable. The matrix could then take the form

$$Z_1 = d_1y_1 + d_2y_2 + d_3y_3$$
$$Z_2 = K. \ Z_1$$
$$Z_3 = a_3y_1 + b_3y_2 + c_3y_3$$

where K is a constant, and

$$d_1 = a_1 + ka_2$$
$$d_2 = b_1 + kb_2$$
$$d_3 = c_1 + kc_2$$

4) Various terms in the matrix can be set to zero (i.e. omitted). This can take several forms, such as the omission of single terms, or the omission of a complete row or complete column. The effect will be equivalent to introducing noise into the system.

It is of interest to compare these strategies with the types of behaviour reported by Miller (1962). These were:

1) Omission – the input is ignored and not dealt with.

2) Approximating – the system emits a response that approximates to the desired output.

3) Chunking – similar inputs are grouped together and treated as a unit.

4) Filtering – inputs of lesser importance are not attended to.

There are obvious similarities between the theoretical strategies under overload and the description of systems behaviour provided by Miller. It is of interest that he reports that the most frequent behaviour is omission, corresponding to computing BZt at intervals longer than those required by the rate of change in the input. This suggests that the control mechanism generally does little to change the nature of its model of the system, but perseveres with it. This is obviously true of mechanical or electronic devices, and it has been suggested elsewhere that managerial systems will also have this characteristic under stress.

However, such a strategy implies that the control action will lag further and further behind the input, until eventually (with a pure sine wave input) the system will switch from negative to positive feedback. This transition will be abrupt. With a complex input, the control output will have a lower correlation with the input, and there may be a point at which this correlation falls abruptly in value if the input contains strong periodic components. Thus overall the control index can be expected to follow the general form of Graph 5.30.

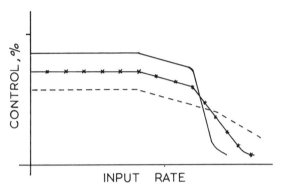

GRAPH 5.30 Theoretical curves for the variation of control with input rate, for sine, periodic and random inputs.

These curves have the general nature of the results reported by Miller (1962). This tends to confirm the importance of speed of data processing to the adequacy of control and hence also confirm the importance of simple models. Although the discussion has been in terms of computable models, the conclusions are also applicable to the non-computable models which have also been described. It is worth reiterating that management models – in the sense of the managers perception of his environment, which he uses in his day-to-day decision making – are likely to be of this latter type. The advantage of this type of modelling is that it can be used in situations that are mathematically intractable; they do not depend upon knowledge of the mechanism whereby variables interact. Although computationally inefficient, they are usable in a much wider variety of situations.

5.4 In Conclusion

The reader may well feel that this book has raised more questions than it has supplied answers. This is not a conclusion with which the author would wish to quarrel. He would, however, proffer a gentle reminder that it was never the intention to provide a set of ready-made solutions. If this book has shown that the workings of management are not wholly mysterious, but can be understood to arise quite rationally from general principles of information processing and control, then it has achieved a major objective.

It is to be hoped, of course, that matters will not rest there. A better understanding is a stepping-stone to progress. And, to reiterate the opening theme, there is much room for improvement in our present institutions.

References

ACKOFF, R.L. and SANSIENI, D.W., *Fundamentals of Operational Research,* John Wiley, 1968.

ALLEN, H.S. and MAXWELL, R.S., *A Text-book of Heat, Part III,* Macmillan, 1952.

ANSOFF, H.I., *Corporate Strategy,* McGraw-Hill, 1965.

ARGYRIS, C., *Personality and Organisation,* Harper and Row, 1957.

ARGYRIS, C., *Understanding Organisational Behaviour,* Tavistock, 1960.

ARGYRIS, C., *Interpersonal Competence and Organisational Effectiveness,* Tavistock, 1962.

ARISTOTLE, *Ethics,* viii, 9.

ASHBY, W.R., *An Introduction to Cybernetics,* John Wiley, 1956.

ASHBY, W.R., *Design for a Brain,* John Wiley, 1960.

ASHBY W.R. and CONANT, R.C., Every Good Regulator of a System Must Be a Model of that System, *Int. J. Systems Sci.* Vol 1, 1970.

BAKKE, E. W., *Bonds of Organisation,* Harper and Row, 1950.

BAKKE, E.W., *The Fusion Process-Interim Report of the Labour and Management Centre,* Yale University Press, 1953.

BAKKE, E.W., 'Concept of the Social Organisation' in *Modern Organisation Theory,* ed. Mason and Hall, Chapman and Hall, 1959.

BARNARD, C., *The Functions of the Executive,* Harvard University Press, 1948.

BARNARD, C., *Organisation and Management,* Harvard University Press, 1948.

BARNES, P., *Company Organisation: Theory and Practice,* Allen and Unwin, 1970.

BEER, S., *Cybernetics and Management,* English University Press, 1959.

BEER, S., 'Towards the Cybernetic Factory' in *Principles of Self-Organisation,* ed. Von Foerster and Zopf, Pergamon, 1962.

BEER, S., *Decision and Control,* John Wiley, 1966.

BEER, S., *Management Science,* Aldus, 1967.

BLAKE, R.L. and MONTON, J.S., *Building a Dynamic Organisation Through Grid Organisation Development,* Addison-Wesley, 1969.

BLAU, P.M. and SCOTT, W.R., *Formal Organisations: a Comparative Approach,* Routledge and Kegan Paul, 1963.

BRECH, E.F.L., *Organisation; the Framework of Management,* Longman Green, 1957.

BRECH, E.F.L., *The Principles and Practice of Management,* Longman Green, 1963.

BROWN, W., *Exploration in Management,* Heinemann, 1960

BROWN, W., *Piecework Abandoned,* Heinemann, 1962.

BROWN, W., *Organisation,* Heinemann, 1971.

BURNHAM, J., *The Managerial Revolution,* Harper and Row, 1941.

BURNS, T. and STALKER, G.M., *The*

Management of Innovation, Tavistock, 1961.

CAMERON, A., *Principles of Management,* Harrap, 1948.

CHOMSKY, N., *Syntactic Structures,* Moulton, 1957.

CUTCLIFFE, F.G.C. and STRANK, R.H.D., *Analysing Catering Operations,* Edward Arnold, 1968.

DEWAN, E.W., *Cybernetics and the Management of Large Systems,* Spartan Books, 1969.

DISTEFANO, J.J. STUBBERUD, A.R. and WILLIAMS, I.J., *Feedback and Control Systems,* McGraw-Hill, 1967.

DONALD, A.G., *Management, Information and Systems,* Pergamon, 1967.

DRUCKER, P.F., *The Future of Industrial Man,* Heinemann, 1943.

DRUCKER, P.F., *Concept of the Corporation,* John Day, 1946.

DRUCKER, P.F., *The End of Economic Man,* Heinemann, 1953.

DRUCKER, P.F., *The Practice of Management,* Mercury, 1955.

DRUCKER, P.F., 'What Communication Means', *Management Today,* March 1970.

DUCKWORTH, E., *A Guide to Operational Research,* Methuen, 1962.

DUERR, C., *Management Kinetics,* McGraw-Hill, 1971.

ECKSTEIN, H., *Pressure Group Politics,* Allen and Unwin, 1969.

EDWARDS, R.S., and TOWNSEND, H., *Studies in Organisation,* Macmillan, 1961.

EILON, S., Management Games, *Op. Res. Quart,* **14,** 2, 1963.

EMERY, F.E., and TRIST, E.L., 'Socio-technical Systems' in *Management Science: Models and Techniques,* ed. Churchman and Verhulst, Pergamon, 1960.

FAYOL, H., *General and Industrial Management,* (trans. Storrs, C.) Pitman, 1949.

FOLLET, M.P., *The New State,* Longman Green, 1920.

FOLLET, M.P., *Creative Experience,* Longman Green, 1924.

FOLLET, M.P., *Dynamic Administration,* Pitman, 1941.

FORRESTER, J.W., *Industrial Dynamics,* Massachusetts Institute of Technology Press, 1961.

FORRESTER, J.W., *Principles of Systems,* Wright Allen, 1968.

FORRESTER, J.W., *Urban Dynamics,* Massachusetts Institute of Technology Press, 1969.

FORRESTER, J.W., *World Dynamics,* Wright Allen, 1971.

GAGNE, R.M., *Psychological Principles In System Development,* Holt, 1962.

GAMSOM, W.A., 'Influence in Use' in *Power and Discontent,* Dorsey Press, 1968.

GARNER, K.C., 'The Evaluation of Human Operator Coupled Dynamic Systems' in *The Human Operator in Complex Systems,* ed. Singleton, W.T., Pergamon, 1968.

GEORGE, F.H., *The Brain as a Computer,* Pergamon, 1961.

GEORGE, F.H., *Cybernetics in Management,* Pan Books, 1970.

GEORGE, F.H., *The Anatomy of Business,* Associated Business Programmes, 1974.

GOODE, H.H., and MACHOL, R.E., *Systems Engineering,* McGraw-Hill, 1957.

GOULDNER, A.W., *Patterns of Industrial Bureaucracy,* Routledge and Kegan Paul, 1955.

GRAICUNAS, V.A., Relationship in Organisation, *Bull. Int. Management Institute,* 1933.

GRAINGER, C.H., *The Hierarchy of Objectives,* Harvard Business Review, May, 1964.

HART, B.L.J., *Dynamic Systems Design,*

Business Publications Limited, 1964.

HERZBERG, F., *Work and the Nature of Man,* World Publications, 1966.

HICK, H.E., On the Rate of Gain of Information, *Q.J. Exp. Psychol.,* 1952.

HORROCKS, B., *A Full Life,* Fontana, 1965.

HUMBLE, J., *Improving Business Results,* McGraw-Hill, 1968.

HUMBLE, J., *Management by Objectives in Action,* McGraw-Hill, 1970.

JANKOWICZ, A., Strategic Management Control, *International Journal of Systems Science,* Vol. I, 1973.

JAQUES, E., *The Changing Culture of a Factory,* Tavistock, 1951.

JAQUES, E., *The Measurement of Responsibility,* Tavistock, 1956.

JAQUES, E., *Equitable Payment,* Heinemann, 1961.

KAUFMAN, A., *The Science of Decision Making,* Weidenfeld Nicholson, 1968.

KEPNER, P. and TREGOE, N., *The Rational Manager,* McGraw-Hill, 1965.

LEE, J., *An Introduction to Industrial Administration,* Pitman, 1925.

LEONTIEFF, W.W., *The Structure of the American Economy, 1919-29,* Harvard, 1941.

LIKERT, R., *New Patterns of Management,* McGraw-Hill, 1961.

LITTERER, J.A., *Organisations: Structure and Behaviour,* John Wiley, 1963.

LUCE, R.D. and RAIFFA, H., *Games and Decisions,* John Wiley, 1957.

MACHIAVELLI, *The Prince,* Penguin, 1961.

MACKAY, D.M., 'Quantal Aspects of Scientific Information', *Phil. Mag.* 41, 1950.

MASLOW, E., *Eupsychian Management,* Dorsey, 1965.

MASLOW, E., *Motivation and Personality,* Harper Row, 1970.

MAYO, E., *The Human Problems of an Industrial Civilisation,* Macmillan, 1933.

MAYO, E., *The Social Problems of an Industrial, Civilisation,* Routledge and Kegan Paul, 1940.

McGREGOR, D., *The Human Side of Enterprise,* McGraw-Hill, 1960.

MILLER, A.K., *The Enterprise and It's Environment,* Tavistock, 1963.

MILLER, A.K., and RICE, A.K., *Systems of Organisation,* Tavistock, 1967.

MILLER, J.G., 'Information Input Overload' in *Self Organising Systems,* ed. Yovits, Jacobi and Cameron, Pergamon, 1962.

MOONEY, J.D. and REILEY, A.C., *Onward Industry,* Harper and Row, 1931.

MORRIS, C.W., *Signs, Language, and Behaviour,* Prentice Hall, 1946.

NETTL, J.P., *Consensus and Elite Domination,* Political Studies Vol. 13 No. 1, 1965.

NEWMAN, D., *Organisation Design,* Arnold, 1973.

NEWMAN, D. and ROWBOTTOM, K.E., *Organisation Analysis,* Heinemann, 1968.

OLSON, M., *The Logic of Collective Action,* Harvard, 1965.

PARKINSON, C.N., *Parkinson's Law,* Murray, 1958.

PASK, A.G., 'Industrial Cybernetics' in *An Approach to Cybernetics,* Radius, 1961.

PASK, A.G., The Cybernetics of Behaviour and Cognition Extending the Meaning of Goal, *Brunel Monograph,* 5, Brunel University Press, 1969.

PASK, A.G., *Tutorial Communication,* 1971.

PFIFFNER, J.M. and SHERWOOD, F.P., *Administrative Organisation,* Prentice-Hall, 1960.

PLATO, *The Republic,* Books II and VIII.

PORTER, B., Probability of Stability of Complex Systems, *International Journal of Systems Science,* 3.1., 1972.

RIVETT, B., *Concepts in Operational*

Research, Watts, 1968.

SHANNON, C.E. and WEAVER, W., *The Mathematical Theory of Communication,* University of Illinois Press, 1949.

SHELDON, O., *The Philosophy of Management,* Pitman, 1924.

SIMON, H.A. and MARCH, J.G., *Organisations,* John Wiley, 1958.

SIMON, H.A., *Administrative Behaviour,* Macmillan, 1960.

SIMON, H.A., *The New Science of Management Decision,* Harper and Row, 1960.

SIMON, H.A., Theories of Decision-Making in Economics and Management Science', *American Economic Review,* Vol. 3, 1959.

SIMON, H.A., On the Concept of Organisational Goal, *Administration Science Quarterly,* Vol. 9, 1964.

SOMMNERHOFF, G., *Analytical Biology,* Oxford University Press, 1950.

STEWART, R., *The Reality of Management,* Heinemann, 1963.

STRANK, R.H.D., *Ergonomics: Functional Design for the Catering Industry,* Edward Arnold, 1971.

TAYLOR, F.W., *Shop Management,* Harper and Row, 1903.

TAYLOR, F.W., *Principles of Scientific Management,* Harper and Row, 1911.

TAYLOR, F.W., *Scientific Management,* Harper and Row, 1947.

TOWNSEND, R., *Up the Organisation,* Michael Joseph, 1970.

TUSTIN, A., *The Mechanism of Economic System,* Heinemann, 1953.

URWICK, L., *The Elements of Administration,* Pitman, 1947.

URWICK, L., *Notes on the Theory of Organisation,* American Management Association, 1952.

URWICK, L. and BRECH, E.F.L., *The Making of Scientific Management,* Pitman, 1950.

VON BERTALANFFY, L., *General System Theory,* General System Yearbook Vol. 1, 1956.

WEBER, M., *The Protestant Ethic and the Spirit of Capitalism,* Allen and Unwin, 1930.

WEBER, M., *The Theory of Social and Economic Organisation,* Free Press, 1947.

WELFORD, A.T., *Fundamentals of Skill,* Methuen, 1968.

WHYTE, W.H., *The Organisation Man,* Simon and Shuster, 1956.

WIENER, N., *Cybernetics,* John Wiley, 1948.

WIENER, N., *The Human Use of Human Beings,* Houghton Mifflin, 1950.

WILLIAMS, J.D., *The Compleat Strategist,* McGraw-Hill, 1966.

WOODWARD, J., 'Management and Technology', *Problems of Progress in Industry,* No. 3, HMSO, 1958.

Index

Algorithm, 30
Argyris, 32, 58
Ashby, 40, 45, 65, 75, 122
Authoritarian, 18
Authority, 11, 16, 35

Bakke, 4, 24
Barnard, 4, 22, 58
Beer, 45
Black box, 49, 67, 72, 82, 123, 127
Brain, 46, 80
Brown, 17, 22
Burnham, 35
Burns and Stalker, 9, 83

Cameron, 7
Channel Capacity, 47, 80, 90, 97, 136
Communication, 6, 9, 18, 19, 22, 63, 64, 70
Control, 6, 25, 54, 68
Control index, 92, 94, 95, 101, 116

Decisions, 30, 40, 67
Drucker, 35
Duerr, 17

Economics, 36
Emery and Trist, 10
Entropy, 44, 48, 49
Environment, 18, 19, 34, 53, 83

Fayol, 10, 21
Feedback, 39, 41, 53, 54, 56, 61, 90, 99
Feedforward, 75
Follet, 28
Forecasting in feedback, 107, 114
Functional management, 12
Functionalism, 12, 16

George, 40
Goal, 39, 40, 63, 67, 99
Goal set, 44, 57, 60, 68
Gouldner, 8
Graicunas, 14
Grainger, 5
Grey box, 85, 127

Hawthorne studies, 31
Heuristic, 30
Hierarchy, 13, 17, 23, 55, 73
Humble, 5, 39

Industrial psychology, 22, 70
Informal working group, 31, 70, 82
Information systems, 17, 40
Information theory, 39, 44, 48, 49, 64, 90

Jankowicz, 41, 75
Jaques, 31
Job enlargement, 32

Lee, 14
Likert, 34

Management
    parallel, 69, 71
    serial, 69, 71
Management functions, 21, 33, 45, 53, 55
Management principles, 10, 12, 27,
Mayo, 31
McGregor, 33, 59
Model, 65, 66, 72, 76, 103, 123
Modelling capacity, 81

Neural net, 40, 46
Newman, 17

143

144 MANAGEMENT PRINCIPLES AND PRACTICE

Operational field, 24
Organisation definitions, 4
Organisation types
    bureaucracy, 7, 8
    charismatic, 7
    composite, 10
    conventional, 10
    executive system, 23
    informal, 22, 70, 82
    legislative, 23
    mechanistic, 9
    military, 19, 64
    organic, 9
    representative system, 23
    traditional, 7
Oscillation, 105, 116

P-approach, 59, 71
Parliament, 18, 125
Pask, 39, 63, 70
Psychology, 29, 32
Purposes, 4, 22

Q-approach, 59, 71

Requisite variety, law of, 48, 80
Role system, 23

Scientific management, 27
Self-financing, 18
Self-organisation, 44, 71, 73
Simon, 30, 40
Span of control, 9, 14
Speed of response, 79, 130
Stability, 9, 47, 50, 61, 122
Strategic control, 42, 53, 75, 79
System models, 52, 69
Systems theory, 38

Task allocation, 7, 10
Taylor, 12, 27
Theory X, 33
Theory Y, 33
Time-span of discretion, 31
Trade unions, 4, 16, 18, 23, 35, 60, 71

Urwick, 12, 30

Variety, 48

Weber, 7
Whyte, 36
Woodward, 8, 17, 20